# Systems Biology

Leszek Konieczny •
Irena Roterman-Konieczna • Paweł Spólnik

# Systems Biology

## Functional Strategies of Living Organisms

Second Edition

 Springer

Leszek Konieczny
Chair of Biochemistry
Jagiellonian University Medical College
Krakow, Poland

Irena Roterman-Konieczna
Department of Bioinformatics and Telemedicine
Jagiellonian University Medical College
Krakow, Poland

Paweł Spólnik
KMG Kliniken SE
Wittstock, Germany

This work was supported by Leszek Konieczny and Irena Roterman-Konieczna, Krakow

ISBN 978-3-031-31559-6        ISBN 978-3-031-31557-2    (eBook)
https://doi.org/10.1007/978-3-031-31557-2

Translated from Polish by Piotr Nowakowski. Graphics performed by Romuald Bolesławski.

This Springer imprint is published by the registered company Springer Nature Switzerland AG
The registered company address is: Gewerbestrasse 11, 6330 Cham, Switzerland

# Preface

The issues addressed in this work complement the basic biochemistry curriculum.

The authors assume that the reader is already familiar with the material covered in a classic biochemistry course. An integrative approach—which is attempted here—enables the reader to grasp the entirety of the problem domain.

We believe that the proposed systemic approach to biology will prove useful not only for students but also for teaching staff and for all those interested in the general domain of biology.

Our handbook frequently refers to real-world macroscopic models. The purpose of this exercise is to underline the unity of the laws of physics, chemistry, and biology and, at the same time, the clear and obvious nature of certain solutions derived from biochemistry and the macroscopic world.

We believe that high school students, already in possession of a large body of detailed information, should focus on generalizations rather than on encyclopedic knowledge, which necessarily becomes fragmented and selective as the amount of available data increases. In our view the foremost goal of education is to ensure that knowledge can be put to practical use by associating facts and predicting their consequences. This can only be achieved by acquainting students with the rules and mechanisms governing various processes and phenomena. Our work should therefore be viewed in light of the presented goal.

Approaching the subject of biology from the viewpoint of basic scientific knowledge (physics and chemistry) yields a convenient platform to formulate generalizations. This is why we have divided the subject matter of the handbook into five sections: the structure and function of living organisms, the role of energy in biology, the role of information in living organisms, regulatory processes in biological systems, and the modes of cooperation in such systems.

We believe that these generalizations will provide readers—particularly those interested in expanding their knowledge beyond simple academic minima—with exploitable insight in the field of biochemistry. Our aim is also to propose a generic platform for simulation techniques.

It goes without saying that any attempt at generalizing diverse biological phenomena involves the risk of oversimplifications or overstepping the bounds of science. Such threats also apply to our work and the reader should be fully aware of this fact. As stated above, this study is primarily aimed at students and therefore it assumes the form of a handbook. In order to encourage readers to try their own hand at interpreting observable events, the last section of the book contains some unresolved hypotheses dealing with fundamental biological processes and the phenomenon of life itself. These include key problems in medicine and drug research, e.g., protein folding and proteome construction, as well as the challenge of formulating a proper definition of life. We hope that the presented study will encourage readers to try and develop their own approaches to such problems.

The work is supplemented by references, including other handbooks, selected monographs, and periodicals dealing with the basic problems of biochemistry.

Krakow, Poland                                                                      Leszek Konieczny

# Preface to Second Edition

This second edition of Systems Biology builds upon the original edition which appeared in 2014. It incorporates new insight which has emerged over recent years, including—among others—the process of neoplastic transformation, which is a key topic in medical science, and which has seen an explosion in research driven by advances in genetic engineering, bioengineering and effective applications of CRISPR-Cas9 technology, derived from the bacterial world.

We maintain the original layout and categorization of the subject matter as in the first edition—including the division into five fields, i.e., (1) Structure and function, (2) Energy, (3) Information, (4) Regulation and (5) Co-action. In making this distinction we were guided by the basic concepts in physics in chemistry, upon which biological systems are based. Referring to theoretical foundations promotes an integrative mindset and enables us to present our arguments in flexible matter, with a unified overall vision. Rapid expansion of knowledge means that it is no longer possible to reconcile the classical, descriptive approach to teaching with the constraints of the academic curriculum. Thus, we need to seek alternatives by adopting a systematic approach to the study subject—and the presented work represents an attempt to achieve this goal.

This second edition of our book is supplemented by tools which enable readers to personally experiment with computations discussed in the Information and Regulation chapters. In the former case, the provided software can simulate ways in which various kinds of processes converge upon a common goal. In the case of Regulation, the attached application provides insight into the operation of negative feedback loops: the Reader may introduce changes and observe their effect on the system being simulated.

Information theory is becoming increasingly important as a source of support for biological sciences. Based on its concepts, we can show that the human DNA contains surprisingly little information—approximately 1 GB. This is relatively modest amount of input results in a human being—the most complex system on the Planet. We can therefore conclude that it is not so much of *quantity* of information that matters, as the means by which the available information is processed. This, in turn, leads us to conclude that a modern biology course should acknowledge the causes, mechanisms, and effects of processing information in the scope of biological systems. Such processing follows a distinct hierarchy, which is embodied by the concepts of Systems Biology.

Krakow, Poland                                                        Leszek Konieczny

# Acknowledgments

This publication is partly supported by the European Union's Horizon 2020 research and innovation programme under grant agreement Sano No 857533 and the International Research Agendas programme of the Foundation for Polish Science, co-financed by the European Union under the European Regional Development Fund.

# Introduction

If we are to assume that biology is not subject to its special, unique laws, but rather conforms to the established principles of physics and chemistry, it follows that the biological world consists of self-managing and self-organizing systems which owe their existence to a steady supply of energy and information.

Thermodynamics introduces a distinction between open and closed systems. Reversible processes occurring in closed systems (i.e., independent of their environment) automatically gravitate toward a state of equilibrium which is reached once the velocity of a given reaction in both directions becomes equal. When this balance is achieved, we can say that the reaction has effectively ceased. In a living cell, a similar condition occurs upon death.

Life relies on certain spontaneous processes acting to unbalance the equilibrium. Such processes can only take place when substrates and products of reactions are traded with the environment, i.e., they are only possible in open systems. In turn, achieving a stable level of activity in an open system calls for regulatory mechanisms.

When the reaction consumes or produces resources that are exchanged with the outside world at an uneven rate, the stability criterion can only be satisfied via a negative feedback loop (Fig. 1).

As cells and living organisms are thermodynamically open systems (namely, they correspond to the description presented above), their internal processes must be subject to automatic regulation if balance is to be maintained. Thus, all structures which play a role in balanced biological activity may be treated as components of a feedback loop. This observation enables us to link and integrate seemingly unrelated biological processes.

In light of the above, the biological structures most directly involved in the functions and mechanisms of life can be divided into receptors, effectors, information conduits and elements subject to regulation (reaction products and action results). Exchanging these elements with the environment requires an inflow of energy. Thus, living cells are—by their nature—open systems, requiring an energy source, i.e., a highly exergonic process. They must also possess the ability to exploit their net energy gains and access stores of information, particularly genetic information.

**Fig. 1** Model approach: (**a**) Equilibrium in a closed system; (**b**) steady state in an open system—equal inflow and outflow rates, (**c**) steady state in an open system, automatically regulated via a negative feedback loop, permitting variations in inflow and outflow rates

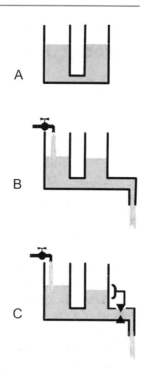

A thermodynamically open system lacking equilibrium due to a steady inflow of energy in the presence of automatic regulation is therefore a good theoretical model of a living organism. We can make a reasonably confident claim that the external signs of life exhibited by a cell reflect its automatic regulatory processes (Fig. 2).

Even for a cell which does not undergo differentiation or division, observing its controlled exchange of substances with the outside environment allows us to conclude that it is, in fact, alive. The additional potential for division and differentiation enables cells to participate in an organized system which is colloquially termed *nature*.

Pursuing growth and adapting to changing environmental conditions calls for specialization which comes at the expense of reduced universality. A specialized cell is no longer self-sufficient. As a consequence, a need for higher forms of intercellular organization emerges. The structure which provides cells with suitable protection and ensures continued homeostasis is called an organism.

In order to explain why organisms exist and how they relate to cells, we can draw analogies between biology and the state-citizen model. Both systems introduce a hierarchy of needs which subordinates the latter to the former. In a biological system subordination applies to cells, while in a state it affects individuals.

Similarly to cells, organisms are subject to automatic regulation. Reaching a higher level of complexity does not imply abandoning automatic regulative

**Fig. 2** Symbolic comparison
between an automaton and a
living organism

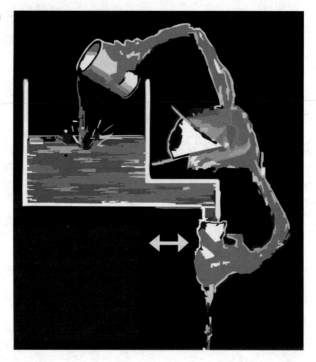

processes in favor of a different means of organization, although signals which reach
individual cells are—for obvious reasons—inherently coercive, forcing the cell to
surrender its independence for the benefit of the organism. This relation is vital for
both the organism and the cell, and therefore underpins the evolution of all biological
structures.

# Contents

# The Structure and Function of Living Organisms

**1**

**Abstract**

The systemic approach adopted in this work focuses on generic biological phenomena. The goal is to explain the strategies employed by biological systems by invoking fundamental concepts in physics, chemistry, and information theory. **Structure and function** is the first of five chapters which make up the book.

In biology, structure and function are tightly interwoven. This phenomenon is closely associated with the principles of evolution. Evolutionary development has produced structures which enable organisms to develop and maintain its architecture, perform actions, and store the resources needed to survive. For this reason we introduce a distinction between support structures (which are akin to construction materials), function-related structures (fulfilling the role of tools and machines), and storage structures (needed to store important substances, achieving a compromise between tight packing and ease of access).

Arguably the most important property of biological structures is their capacity for **self-organization**. Ongoing evolution has led to the emergence of biological constructs capable of exploiting basic physico-chemical interactions in support of directed organization processes, limiting in this way the quantity of the encoded information needed to perform their function. Examples of self-organization include the spontaneous folding in water of polypeptide chains and creation of cellular membranes.

**Keywords**

Self-organization · Support structures · Function-related structures · Storage structures

In biology, structure dictates function.

1. Role and importance of polymers in biology
2. What is meant by "autonomous operation" and how is it achieved?
3. Reasons behind the diversity of intra- and extracellular scaffold structures
4. Role of the cellular membrane as the boundary between the cell and its environment
5. How is the cellular membrane stabilized?
6. Role of polysaccharides in biology—reasons behind the observed diversity
7. Tools and machines in biology—what are the advantages of constructing machines?
8. What does the Michaelis-Menten equation tell us about the properties of enzymes?
9. Mechanism of action of enzyme inhibitors (versus drugs)
10. Types of proteins—structural characteristics and relationship between structure and function
11. How is energy stored in biological systems?
12. Self-organization—its properties and importance in biology
13. How is information stored in biological systems?

The mechanisms described in this chapter can be experimented with using two web applications we provide for the reader's convenience.

HPHOB online tool provides the ability to compute FOD model parameters for arbitrary protein structures. The tool is available at https://hphob.sano.science.

The reliance of nature on self-regulation and self-organization implies a certain specificity in interactions. At the same time, evolutionary pressure acts to eliminate burdensome and disadvantageous structures. Tight coupling between structure and function is therefore ubiquitous in biology. Analysis of the available information points to strategies for creation and exploitation of structures which facilitate certain processes in living organisms.

## 1.1    General Physiochemical Properties of Biological Structures

### 1.1.1    Small-Molecule Structures and Polymers

Biology makes an extensive use of small-molecule structures and polymers. The physical properties of polymer chains make them a key building block in biological structures. There are several reasons as to why polymers are indispensable in nature:

A. Polymers enable multipoint surface contact, which strengthens reactions and yields strong complexes held together with noncovalent bonds. The structure of a polymer makes noncovalent bonding (which is far more common than its

**Fig. 1.1** The effect of noncovalent interaction on the properties of small-molecule compounds and polymers. Noncovalent bonding of (**a**) monomeric units (kinetic energy exceeds compound stabilization energy— compound not formed), (**b**) polymers (noncovalent bonding energy exceeds the kinetic energy of individual interacting compounds (aggregate formed))

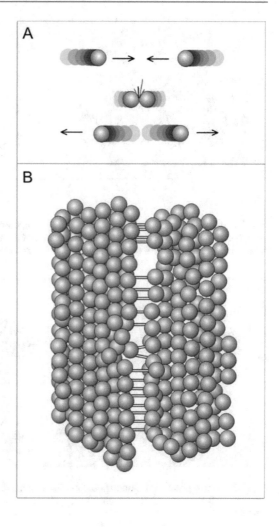

covalent counterpart) a meaningful mode of interaction, sufficient to stabilize compounds which could not otherwise be constructed from monomers.

Figure 1.1 depicts the interaction of monomeric units and polymers in a model system.

B. They enable the use of globular polymer structures as carriers of atypical environmental conditions, particularly apolar microenvironments, isolated from water by a polymer layer. This phenomenon gives rise to active protein compounds, which—owing to the apolar nature of their environment—catalyze many reactions otherwise unattainable in water (Fig. 1.2).

C. The multifocality of interactions involving polymers provides for the emergence of structures unique in terms of their surface shape and orientation (Fig. 1.3). In this way polymers can be used to encode information.

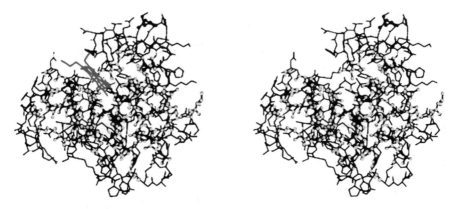

**Fig. 1.2** Distribution of apolar amino acids in a myoglobin molecule (wireframe model). Apolar amino acids are represented as colored spheres. The globin molecules with (left) and without (right) heme were put together to expose the active region (data obtained from the Protein Data Bank)

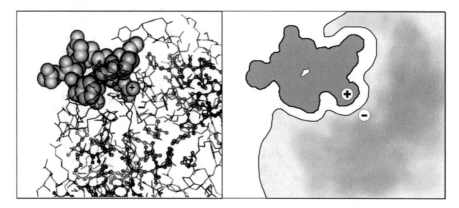

**Fig. 1.3** Active group of immunoglobulin with an attached ligand (polypeptide chain in space-filling model) and a planar cross-section of the same complex (data obtained from the Protein Data Bank)

D. Creation of complexes using noncovalent bonds enables reversible and controllable reactions which are fundamentally important in biological systems.

## 1.1.2   The Biological Purpose of Cellular and Organism Structures

In general terms, biological structures can be divided into three groups, each serving a different purpose:

1. *Supporting structures*, not directly involved in processes which constitute life, but providing a framework and foundation for such processes. These structures

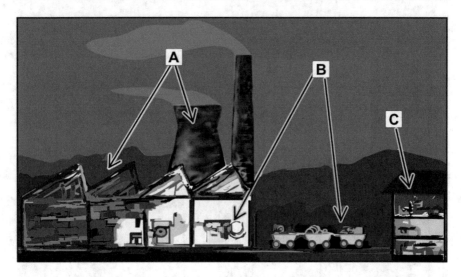

**Fig. 1.4** The factory as an autonomous entity with subunits which fulfill (**a**) supporting, (**b**) functional, and (**c**) storage tasks

perform shielding, reinforcing, and dividing functions, which roughly equate to the role of architecture in our macroscopic world.

2. *Structures directly involved in the functioning of living organisms*, i.e., those that *facilitate metabolism and motion*. If we accept the proposition that biological functions are inseparably tied to regulatory processes and that all functional structures belong to regulatory loops, these structures can be further subdivided into receptors, effectors, and information conduits.

3. *Storage structures*, i.e., structures used to store *information and energy*. Their goal is to fit as much energy (or information) into as small a volume as possible, as well as to enable rapid access to both.

The presented classification acknowledges the divergence of structural elements which form living organisms. The properties of structures belonging to each group are a clear consequence of their evolutionary goals. Similar models exist in many non-biological systems where individual objects interact to perform a common task. An example of such a system is depicted in Fig. 1.4.

It can be noted that the challenges that need to be overcome when, e.g., erecting buildings (such as factories) are substantially different from those involving machines and tools. Furthermore, any production process (and particularly one that is subject to fluctuations in the supply of raw resources and demand for finished goods) must take into account storage capabilities: the shape of manufactured goods should enable efficient storage and easy access.

Figure 1.5 summarizes the properties of the presented structures and the reasons behind introducing this classification.

## BIOLOGICAL STRUCTURES – MOLECULAR LEVEL

| | I. SUPPORTING STRUCTURES | II. STRUCTURES CONNECTED WITH FUNCTION | III. STRUCTURES CONNECTED WITH STORAGE |
|---|---|---|---|
| ORIENTATION TO PROBLEM | Physical strenghtening<br>Bordering | Specific interaction – binding of ligands<br>– complexation of polymers<br>– material transformation | Structures facilitating packing and unwrapping |
| STRUCTURAL CHARACTERISTICS | One or two dimension structures<br>(fibrillar forms, plater) | Globular proteins or huge combined structures.<br>Binding site as the characteristic element | Structures of low solubility facilitating packing of the material through polymerization or association |
| KEY STRUCTURAL DEVICES | Skeleton and cover structures | Receptor structures<br>Effector structures<br>Structures connected with signaling<br>Regulated component of processes | Structures connected with<br>– information<br>– energy<br>– other processes |
| THE DEGREE OF ORGANIZATION | Relatively low<br>Polymers and associates of moderately low structural differentiation | Divers:<br>– Receptor structure: simple or complex<br>– Effector structures: simple (tools)<br>complex (machinery, engines, functioning systems) | High – for information storage (genome)<br>Low – for energy storage (glycogen, fat) |
| BASIC MATERIAL | Proteins, carbohydrates, lipid components characteristic for membranes | Proteins, RNA: active material<br>Other components: passive material | DNA, carbohydrates, fat, iron |

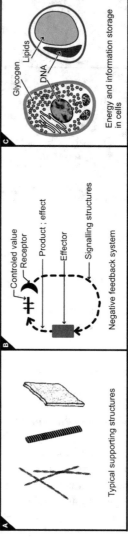

Typical supporting structures

Negative feedback system

Energy and information storage in cells

**Fig. 1.5**  Task-oriented properties of biological structures

### 1.1.3   Supporting Structures

The purpose of these structures is to provide shielding and support and also—owing to their physical properties—ensure sufficient rigidity and elasticity of biological entities. Supporting elements are, by their nature, one- or two-dimensional: they include fibers, membranes, and walls.

As conditions and processes which take place inside a cell differ significantly from those encountered outside, supporting structures must be suitably differentiated to suit both. Thus, each cell and each organism make use of many different types of supporting structures.

#### 1.1.3.1 Cellular Supporting Structures

Cells owe their existence to fibers forming the so-called cytoskeletons. The cytoskeleton can best be described as an aggregation of structures which provide the cell with rigidity while also ensuring certain elasticity, similar to the scaffold of a tent. However, unlike a manmade scaffold, the cytoskeleton must allow for changes associated with the biological function of the cell (relative motion of cellular organelles; motion of the entire cell). Motion is sustained by maintaining a dynamic balance between formation and decomposition of fibers. In addition to directed motion, the supporting structure must also enable expansion and contraction of the cell.

As explained above, motion is facilitated by fibrillar structures which consist of globular subcomponents, attached and detached to one another in a steady process of synthesis and decomposition. Typical fibrillar structures which fulfill these criteria include microtubules and microfilaments (Fig. 1.6 A and B).

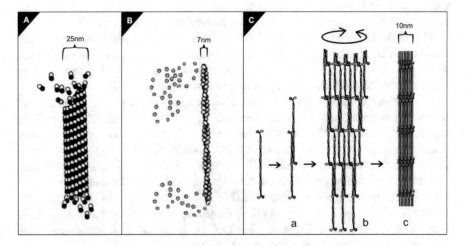

**Fig. 1.6** Basic components of cytoskeletons: (**A**) microtubules, (**B**) microfilaments, and (**C**) intermediate filaments: (a) dimeric forms, (b) tetrameric forms, and (c) final form of the fiber

a                          b

**Fig. 1.7** Smokestack (**a**) and rope (**b**) as physical models of supporting structures shaping the mechanical properties of cytoskeletons

Microtubules are pipe-like structures (Fig. 1.6 A) characterized by significant rigidity and ability to withstand mechanical forces. In terms of their structural properties, they can best be compared to a factory smokestack (Fig. 1.7).

The chain-like microfibrils (Fig. 1.6 B) also consist of globular units. Their elasticity is far greater than that of microtubules. In a way, they resemble ropes, although their tensile strength is not as great (a consequence of the fact that, just like microtubules, they are also composed of globular units).

In contrast, structures formed from fibrillar compounds are far stronger and capable of resisting significant longitudinal forces. The most distinctive cellular supporting structure is the so-called intermediate filament—another type of fiber which contributes to the scaffolding of a cell. Intermediate filaments differ significantly from the structures mentioned above. Their monomeric elements are fibrillar, unlike microtubules and microfilaments which consist of globular units (Fig. 1.6 C). They assume the form of parallel dimers (Fig. 1.6C.a) forming a structure which can be compared to roof tiling (Fig. 1.6C.b) Together, these units constitute the final form of the intermediate filament (Fig. 1.6C.c). Intermediate filaments resemble ropes with respect to their physical properties (Fig. 1.7).

**Fig. 1.8** The tent as a support model, comprising elements with varying rigidity and elasticity (poles and guy ropes), which together ensure proper shape of the finished structure

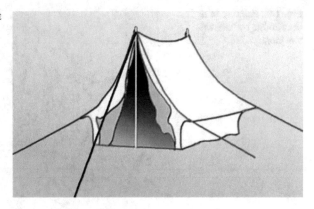

Unlike microtubules and microfilaments, intermediate filaments are apolar. Their dimers are bound in a counter-parallel fashion, such that their terminal carboxyl and amine domains enter a 1:1 relation with one another. Moreover, it seems that intermediate filaments are not autonomous in determining their direction of expansion within a cell and that they are assisted in this respect by microtubules. Research suggests that intermediate filaments do not degrade in the same way as microtubules: instead of yielding a substrate which can be further used to construct supramolecular structures, they are instead digested and broken down into fragments. Owing to these properties, intermediate filaments are more closely related to extracellular supporting structures. They are not restricted to acting within the boundaries of a cell; instead, they may permeate the cellular membrane (at points where desmosomes are located) and attach themselves to the cytoskeletons of neighboring cells. They stabilize multicellular complexes in epithelial tissues and organs.

Elongation of intermediate filaments is made possible by specific interactions of globular fragments positioned at the ends of monomeric fibrils (Fig. 1.6C). In addition to elongation, the filaments are also capable of growing in thickness. The fibrous nature of their monomeric components grants them physical resilience and supports their biological function as supporting structures, although their relative lack of autonomous direction and low renewal rate make them the most static component of the cellular cytoskeleton.

Intermediate filaments are nonuniform in nature, though they all share a similar basic structure. Their lack of uniformity is a result of specific adaptation to the requirements of various types of cells.

As structures, microfilaments and microtubules are not strictly limited to facilitating motion: their rigidity and elasticity enable them to play an important role in reinforcing the structure of the cell.

It appears that microfilaments concentrated near the boundaries of the cell are responsible for maintaining proper tension of its membrane, while microtubules provide the cell with appropriate rigidity (Fig. 1.8).

Observing the dynamic changes in the length of microfilaments and microtubules as well as the motion patterns effected by interaction with motor proteins enables us

to classify such cytoskeletal structures as *functional* in nature. This association of supporting structures with motion is quite peculiar and unlike most manmade objects. In search for suitable macroscopic equivalents, we can refer to escalators, rotating stages, and drawbridges, which—depending on the situation—may serve as barriers or roads (Fig. 1.9).

### 1.1.3.2 Cellular Shielding Structures

The second class of cellular supporting structures comprises of shielding structures—specifically, cellular membranes (consisting of phospholipids) and extracellular membranes (e.g., the basal membrane, composed of fibrillar protein and glycoprotein aggregates). Phospholipid cellular membranes also include glycolipids, cholesterol, and various proteins. While all integrated proteins contribute to the stability of their membranes, only some of them (such as spectrin) are clearly supportive in nature, serving as a scaffold for the membrane itself.

The cohesiveness and rigidity of membranes are important in determining their function. Phospholipid solutions act as liquid crystals: in water they form a planar (non-spherical) micelle capable of performing the functions of a membrane. Such interlocking panel-like arrangement is made possible by minimizing repulsive forces through mutual cancellation of positive (choline) and negative (phosphoric acid) charges in the polar fragments of phospholipid molecules (Fig. 1.10).

The lipotropic and liquid crystal properties of membranes enable them to retain some characteristics of a liquid. At normal body temperatures, cellular membranes have the approximate thickness of tar. The elasticity of the membrane can be increased or decreased without affecting ambient temperature, by altering the ratio of saturated and unsaturated fatty acids (Fig. 1.11).

The presence of a double bond and the resulting divergence of *cis-* and *trans-*configurations reduce adhesion and inhibit molecule interactions, thus lowering the

**Fig. 1.10** Close internal and external placement of positive and negative charges in the polar layer of the phospholipid membrane facilitates interlocking arrangement of molecules, resulting in the formation of a planar micelle (neutralization of charges eliminates repulsive forces)

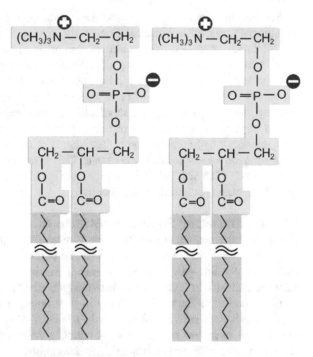

**Fig. 1.11** The impact of double bonds and *cis-* and *trans*-isomers on the area of adhesion and range of potential molecular interactions in the lipid layer of a cellular membrane

overall rigidity of the system. On the other hand, the increased adhesion of saturated molecules results in increased rigidity, more closely resembling that of a solid. An identical phenomenon occurs in margarine production where the viscosity of vegetable oils is increased via reduction reactions.

Another important mechanism which contributes to the rigidity of cellular membranes is cholesterol intercalation.

Cellular membranes (both cytoplasmic and nuclear) may be further stabilized by a two-dimensional lattice consisting of fibrillar proteins parallel to the membrane. Such a scaffold is formed, e.g., by spectrin molecules in the membranes of erythrocytes and other types of cells. Spectrin is a protein-based fibrillar complex made up of a coiled pair of giant polypeptide chains designated $\alpha$ and $\beta$. Both chains consist of periodic sequences of repeating protein domains, each in the shape of a helix, separated by hinge fragments. Together with microfibrils, they create a lattice attached to the underside of the membrane. This structure is particularly important in erythrocytes which lack the skeleton present in eukaryotic cells and therefore require special membrane stabilization mechanisms.

### 1.1.3.3 Extracellular Supporting Structures

Similar to intermediate filaments, protein microfilaments present in extracellular supporting structures are not directly involved in motion; thus they may be fibrillar, with relatively limited exchange dynamics. Examples of such structures include collagen and elastin. Just like in intermediate filaments, their fibrillar nature is primarily due to the lack of variability in their amino acid composition (repeating GLY-PRO-X sequences in collagen and PRO-GLY-VAL-GLY in elastin).

Contrary to amino acid sequences which give rise to fibrillar structures, the globular nature of functional proteins is determined by high variability of amino acid chains, in particular the uneven distribution of hydrophobic residues.

The abundance of proline and glycine in collagen chains determines their structure (although it should be noted that collagen owes its fibrillar nature chiefly to the repeatability of amino acid sequences). A high number of proline and glycine units impose specific torsion on the resulting spiral, while their relatively small volumes (particularly in the case of glycine) provide the resulting strands with high cohesiveness which translates into increased rigidity.

Elastin is a special protein responsible for the elasticity of tissues. The very notion of elasticity is somewhat unusual as the material in question must be capable of fluid noncooperative transition from a folded to a stretched state, yet retain the tendency to return to the initial folded state (much like rubber). Such properties are not found in secondary and tertiary structures of polypeptide chains, which—being cooperative—tend to unfold in an abrupt manner, kinetically equivalent to the phase transition observed in denaturation (Fig. 1.12).

The elasticity of elastin results from shearing hydrogen bonds. In this respect, it may be compared to mechanically induced melting. The fluid nature of this process proves that the structure of elastin peptide chains is highly random (Fig. 1.13).

Owing to the rotational freedom of glycine in polypeptide chains and the fact that glycine constitutes 50% of elastin, the resulting protein may assume a practically unlimited number of folding configurations. The force which induces folding is most likely associated with hydrogen bonds stabilizing the so-called $\beta$-turns, wherever proline and glycine are directly adjacent. In addition, the hydrophobic nature of elastin (a consequence of valine abundance) protects hydrogen bonds from coming

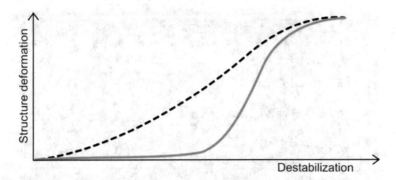

**Fig. 1.12** Transitions in a cooperative (solid line) and noncooperative (dashed line) system as a response to destabilizing forces

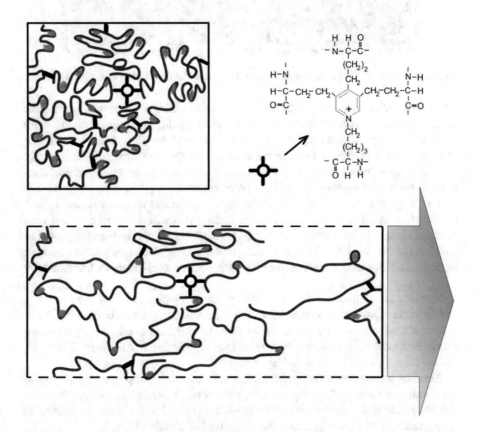

**Fig. 1.13** A hypothetical model explaining the elasticity and noncooperative melting of elastin. The diagram in the top right-hand corner presents the symbol and structure of desmosine. Darkened areas indicate potential hydrogen bonding sites

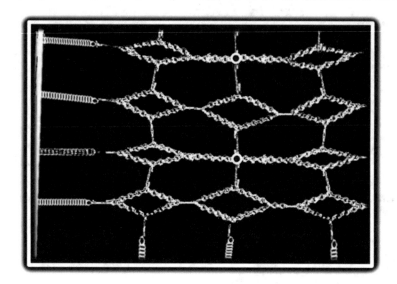

**Fig. 1.14** Fragment of a spring mattress—a mechanical model functionally analogous to elastin

into contact with water, thus facilitating smooth refolding with many intermediate states and high structural randomness. Covalent binding of peptide chains with desmosine bridges ensures continuity of the resulting material, enabling it to assume one-, two-, and even three-dimensional shapes. This type of structure is functionally equivalent to a spring mattress, whose individual components are capable of contraction yet bound together in a stable configuration (Fig. 1.14).

The polypeptide chains of collagen are mostly polar and therefore unable to fold in a random fashion. Here, water-isolated hydrogen bonds can exist only between separate chains; hence collagen stabilizes in the form of tightly fused triple-chain conglomerates (Fig. 1.15). The tight fusing of collagen fibrils, made possible by their high glycine content, increases their rigidity and enables them to mechanically stabilize connective tissue.

Keratins are another family of fibrillar proteins, distinguished by their purpose and function. They act as intermediate filaments and are essentially intracellular in nature, although they may occasionally traverse cellular boundaries. Synthesized inside living cells, they persist through cell death, forming a solid, insoluble mass, expressed, e.g., as fingernails and hair.

Keratins include repeating seven amino acid sequences (a property common in intermediate filaments). This repeatability determines their fibrillar nature, while low polarity and high numbers of disulfide bonds linking individual chains contribute to the mechanical resilience of structures composed of keratin fibers.

**Fig. 1.15** Structure of collagen, showing the location of water-isolated hydrogen bonds which link individual strands

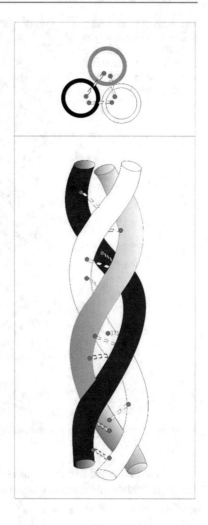

## 1.1.3.4 Polysaccharides as Supporting Structures

A special class of supporting structures consists of polysaccharides. These substances are characterized by high rigidity, which is due to the stabilizing influence of β-glycosidic bonds (Figs. 1.16 and 1.17).

Contrary to storage sugars, i.e., α-type polysaccharides which exhibit the tendency to form helical structures, the presence of β-glycosidic bonds results in threadlike strands which interact with one another in a highly specific fashion. Good contact between individual fibrils, particularly the high number of point-to-point interactions observed in β-type polysaccharide chains, makes the resulting complexes highly resistant to mechanical damage and thus useful in supporting structures. The relatively low diversity of monomeric units observed in hydrocarbon polymers (compared to proteins) results in similarly low diversity among their derivative structures.

**Fig. 1.16** The branching helical structure of α-type polysaccharide chains reduces the potential for inter-chain contact and limits their packing density

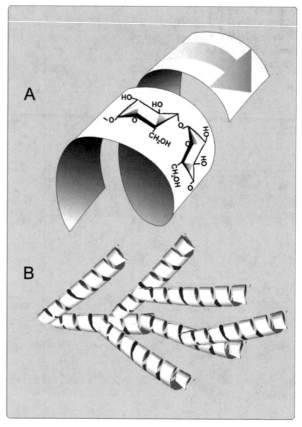

It should be noted that the general diversity of proteins is a direct consequence of the enormous variability in both the number and ordering (sequence) of amino acids, giving rise to a wide array of function-specific structures. By the same token, polysaccharides consisting of a single type of monomer cannot adapt to diverse functions and are therefore better suited to providing structural support.

Cellulose is a conglomerate of long fibrils, each comprising hundreds of glucose molecules. The most frequently observed unit of cellulose is a packet consisting of approximately 60–80 fused threadlike polymer strands. Further packing yields a clustered, uniform mass.

Polysaccharide frameworks are ubiquitous in the realm of plants and among primitive animals such as insects. In the latter group, the most frequently occurring polysaccharide is chitin—an interesting example of how nature reinforces polypeptide structures by increasing the number of inter-chain interactions. In the case of chitin, this function is performed by amide clusters, which easily form hydrogen bonds and therefore increase the overall tensile strength of the framework. On the other hand, rigidity can also be reduced via uniform polarization of individual

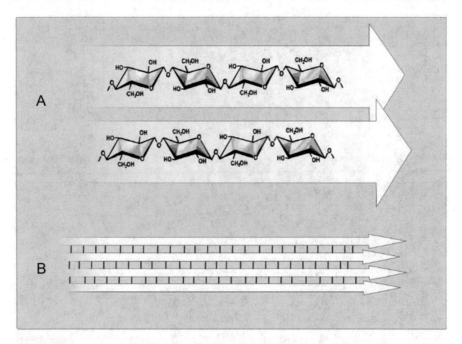

**Fig. 1.17** The simple, nonhelical structure of β-type polysaccharide chains facilitates easy interchain contact and ensures good mechanical resilience of the resulting complexes

monomers comprising the polymer chain, which increases their electrostatic repulsion and lowers the tensile strength of the resulting material (Fig. 1.18).

Polar polysaccharides, such as keratan sulfate or chondroitin sulfate, are encountered in tissues as chains. Together with specific proteins and (optionally) hyaluronic acid, they can form giant, branching, treelike structures which play an important role in binding water molecules. Their interaction with collagen fibers (with which they form complexes) gives the connective tissue the required rigidity and turgor as a filler (Fig. 1.19).

Glycosaminoglycans are a good example of how polar polysaccharides can be employed in the connective tissue (Fig. 1.19).

In summary, we can state that—contrary to the cellular cytoskeleton—extracellular polymer structures (both protein- and carbohydrate-based) play a strictly passive and supportive role in living organisms.

## 1.1.4   Structures Associated with Biological Functions

The term "functional structures" denotes structures which facilitate metabolism and motion (respectively—changes in the properties of matter and changes in location).

Reversible creation of specific complexes enables biological functions.

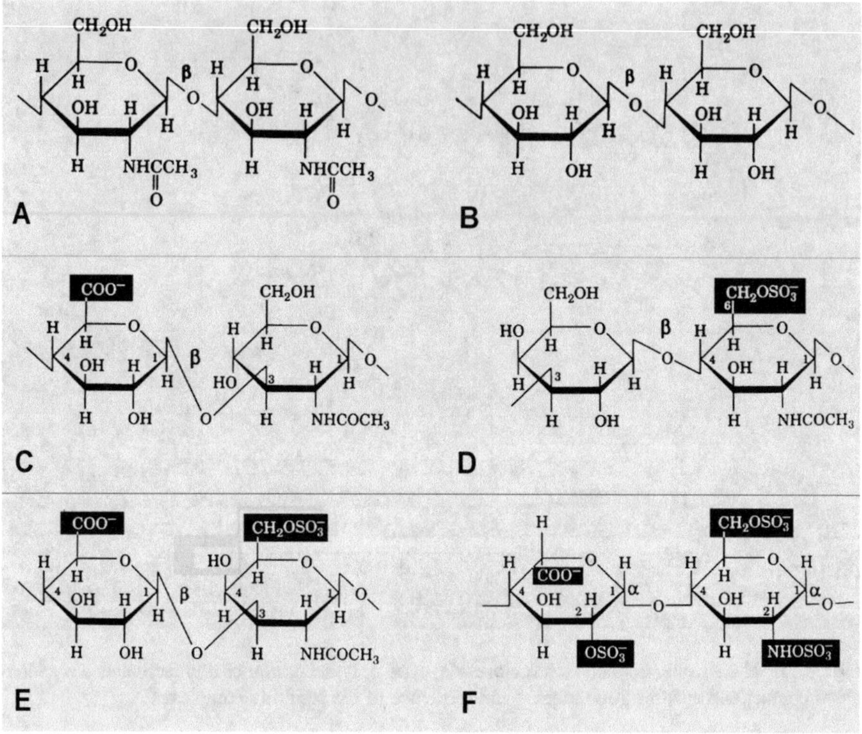

**Fig. 1.18** Fragments of polysaccharides with differing levels of polarity: (**a**) chitin, (**b**) cellulose, (**c**) hyaluronate, (**d**) keratan sulfate, (**e**) chondroitin sulfate, and (**f**) heparin

The seemingly endless diversity of biological functions frustrates all but the most persistent attempts at classification. For the purpose of this handbook, we assume that each function can be associated either with a single cell or with a living organism. In both cases, biological functions are strictly subordinate to automatic regulation, based—in a stable state—on negative feedback loops, and in processes associated with change (for instance, in embryonic development)—on automatic execution of predetermined biological programs. Individual components of a cell cannot perform regulatory functions on their own (just like a thermometer or heat pump extracted from a refrigerator). Thus, each element involved in the biological activity of a cell or organism must necessarily participate in a regulatory loop based on processing information.

In light of this assumption, we can divide all functional structures into the following categories:

1. Receptor structures
2. Effector structures
3. Information carriers
4. Structures which are subject to regulation (regulated process components)

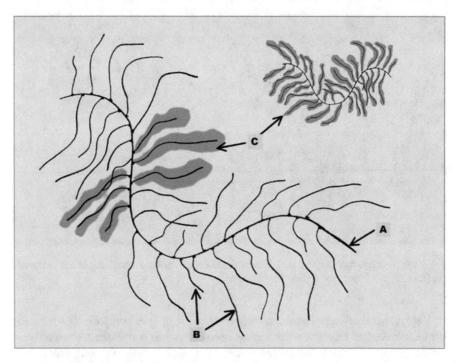

**Fig. 1.19** The characteristic branching structure of proteoglycans: (A) hyaluronic acid, (B) glycoprotein chains, and (C) bound water molecules (shaded areas)

The detector function of a receptor and the action of an effector may be facilitated by simple proteins or by complexes consisting of many protein molecules. Proteins are among the most basic active biological structures. Most of the well-known proteins studied thus far perform effector functions: this group includes enzymes, transport proteins, certain immune system components (complement factors), and myofibrils. Their purpose is to maintain biological systems in a steady state. Our knowledge of receptor structures is somewhat poorer, mostly due to their tight integration with cellular membranes, making them difficult to extract and isolate in a crystalline form.

Involvement in a given function may be either active or passive. Active involvement occurs in proteins and certain forms of RNA. The term "active form" typically denotes the form genetically programmed to create a specific complex and fulfill a specific biological purpose. This form is also responsible for ensuring that the resulting complex is well suited to its function and that it undergoes structural changes while performing that function. Biological functions are usually mediated by proteins. On the other hand, passive involvement in biological mechanisms is characteristic of reaction substrates and products, as well as various messenger molecules such as hormones and DNA.

Contrary to supporting structures, the proteins which participate in functional aspects of life are almost invariably globular.

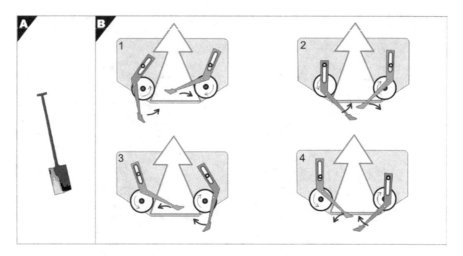

**Fig. 1.20** A simple tool (shovel) (**a**) and an advanced snow collector, consisting of linked shovels (**b**) (four stages of operation depicted for machine **b**)

Receptor and effector structures may be either simple or complex. The need for complex receptors emerges when signals become unclear and difficult to classify. A similar process occurs in effectors, where certain tasks require advanced and well-organized systems. For instance, most enzymatic effectors can be divided into distinct stages (catalysis of complex reactions often involves many separate steps). Such multienzyme effectors assist, e.g., in glycolysis, synthesis of purines and pyrimidines, as well as synthesis and degradation of fatty acids.

Simple structures, including individual enzymes and components of multienzyme systems, can be treated as "tools" available to the cell, while advanced systems, consisting of many mechanically linked tools, resemble machines. The rationale behind constructing machines is obvious: certain processes cannot take place without the aid of complex mechanical devices. Combined action of tools enables efficient processing of materials while reducing the need for immediate access to information. In a machine model, individual elements (tools) perform work under the supervision of integrating mechanisms (Fig. 1.20) such as conveyor belts, chains, gears, etc. In a cell the role of conveyor belts is fulfilled by special protein structures acting as booms or servos. Other more advanced structures can be distinguished as well, facilitating the conversion of energy into structural changes. Such mechanisms are often thermally powered, i.e., their action is determined by the laws of Brownian motion.

Machine-like mechanisms are readily encountered in living cells. A classic example is fatty acid synthesis, performed by dedicated machines called synthases. The process is similar to β-oxidation; however synthesizing a fatty acid molecule carries a net increase in the structural ordering of the product. This property implies the need for energy as well as a source of information which is implicitly contained in the structure of synthase, enabling its enzymes to collaborate on a common task.

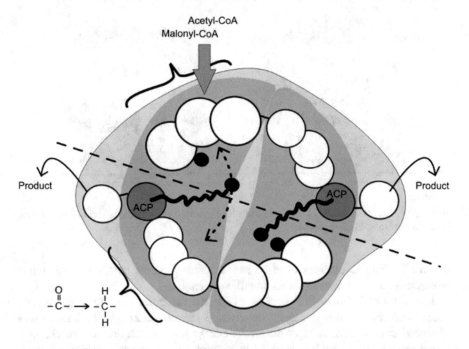

**Fig. 1.21** Schematic model of fatty acid synthase, showing clustered enzymatic domains and servos (ACP). Small black circles represent SH (sulfhydryl) groups, while brackets indicate moieties responsible for specific stages of the synthesis process

In contrast, enzymes charged with degradation of fatty acids require no such information and do not need to form machines as their final product is more chaotic than the substrate they break down.

Fatty acid synthase is a homodimer consisting of two identical (260-kd) subunits which perform enzymatic actions related to the synthesis of palmitic acid. Both structures are cross-linked with loose hinge fragments, enabling relatively unrestricted motion—an important property required for good contact between the enzyme and the chain undergoing synthesis. Owing to the apolarity of the substrate, synthesis occurs in dedicated hydrophobic gaps in which the active moieties are concentrated. The boom-like structure which aligns the substrate with active sites of the complex is called ACP (*acyl carrier protein*). It includes a long, flexible chain, structurally similar to coenzyme A. High flexibility of the whole complex, and particularly of ACP, creates suitable conditions for alignment of the substrate with moieties required at each stage of the synthesis process. Both halves of the machine come into play at certain stages, while their cross-linked nature is primarily a stabilizing factor (Fig. 1.21).

Another example of a biological machine is the pyruvate carboxylase enzyme which consists of two enzymatic domains (biotin carboxylase domain and carboxyl transferase domain) in addition to a biotin-binding domain (Fig. 1.22). Multiunit

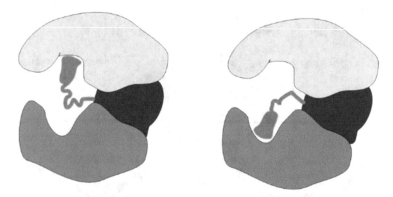

**Fig. 1.22** Functional model of a simple machine (pyruvate carboxylase). The colored section indicates the biotin-binding domain

structures acting as machines can be encountered wherever complex biochemical processes need to be performed in an efficient manner.

As mentioned above, the simplest biological machines employ long, flexible booms and servos with thermally determined motion patterns (Fig. 1.23). More advanced molecular machinery draws the energy for coordination and mechanical action from high-energy bonds. A good example is the ribosome which acts as a shifting matrix for the synthesis of polypeptide chains.

If the purpose of a machine is to generate motion, then a thermally powered machine can accurately be called a motor. This type of action is observed, e.g., in myocytes, where transmission involves reordering of protein structures using the energy generated by hydrolysis of high-energy bonds.

Some production tasks are performed by the entire "factory chains," i.e., structures integrated in cellular membranes in an ordered fashion. This mode of operation is applied in synthesis and distribution of certain proteins (Fig. 1.24). The nucleus of the cell, acting as a "design department," issues blueprints via the "conveyor belt" structure of endoplasmic reticulum. The resulting products enter the Golgi apparatus while undergoing stepwise changes. Inside the apparatus they are "packaged" and cleared for "export" (i.e., expression outside the cell). The "wrapping paper" is actually a special protein called clathrin, whose associations assume the form of vesicles, encapsulating the synthesized "payload" proteins.

In biology, function is generally understood as specific physiochemical action, almost universally mediated by proteins. Most such actions are reversible which means that a single protein molecule may perform its function many times. Functional proteins are usually globular in shape and contain similar types of active sites—pockets or clefts characterized by concentration of exposed hydrophobic residues and appropriate (for a given substrate) arrangement of polar amino acid groups or non-protein-based aggregates, enabling the site to attach to specific types of ligands. In addition, enzymatic active sites include a catalytic element (Fig. 1.25).

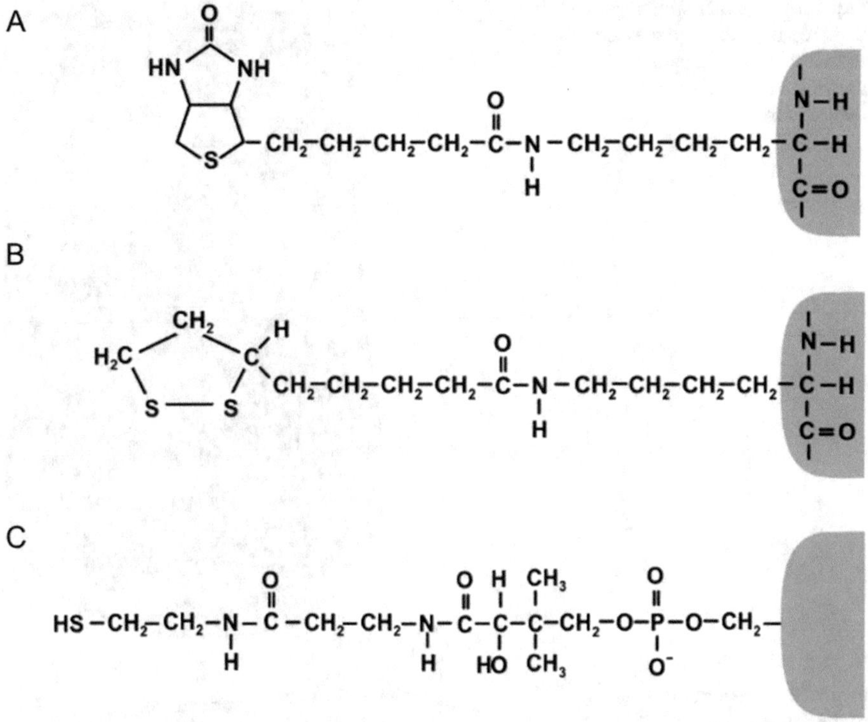

**Fig. 1.23** Examples of exterior area structures: (**a**) biotin-binding domain, (**b**) lipoamide, and (**c**) ACP

This element is directly responsible for catalysis, i.e., for action which affects a specific bond in the substrate and converts it to an intermediate state where it can spontaneously reassemble into the final product. Catalysis can be performed by a special amino acid group, a metal ion, or a coenzyme (Fig. 1.26).

The intermediate state is an unstable form of the substrate, highly susceptible to environmental stimuli. This instability typically hinges on a single molecular bond also. Substrates can enter intermediate states as a result of mechanical stress resulting from local incompatibilities between a tightly bound substrate molecule and the active moiety of the enzyme. A good example is lysozyme whose active moiety strongly binds to polysaccharide fragments consisting of six monomers, one of which is sterically unaligned. The resulting change in the structure of the substrate deprives the glycosidic bond of its stability and eventually leads to hydrolysis.

In general terms, we can state that enzymes accelerate reactions by lowering activation energies for processes which would otherwise occur very slowly or not at all.

Hydrophobic component is required for the formation of stable ligand complexes in aqueous environments, irrespective of structural considerations. This is due to the fact that electrostatic interactions and hydrogen bonds are strongly selected against

**Fig. 1.24** Cellular structures
involved in protein synthesis.
Individual machine parts are
integrated in membranes in an
ordered fashion

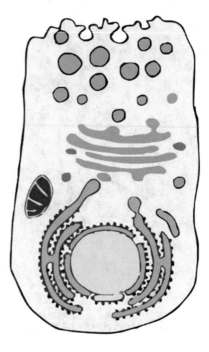

**Fig. 1.25** Schematic diagram
of the active site of trypsin.
The exposed serine oxygen is
the catalytic element
responsible for attacking the
Cα carbon of the amino acid
involved in the peptide bond

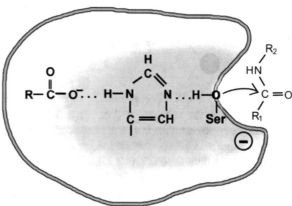

in physiological saline solutions, as water itself is a good donor and acceptor of
protons, while salt ions act as shields. Thus, saline environments limit surface
interactions and prevent native proteins from aggregating even at high
concentrations (for instance, in blood plasma or egg white). Since spontaneous
noncovalent surface interactions are very infrequent, the shape and structure of
active sites—with high concentrations of hydrophobic residues—make them the
preferred area of interaction between functional proteins and their ligands. They
alone provide the appropriate conditions for the formation of hydrogen bonds;
moreover, their structure may determine the specific nature of interaction. The

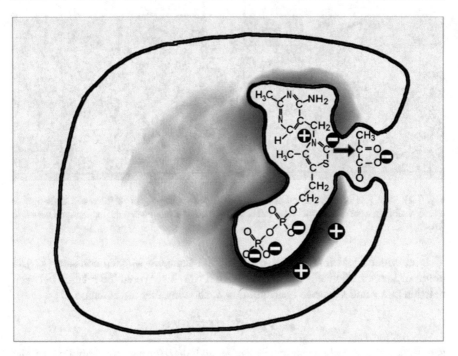

**Fig. 1.26**   Schematic diagram of a protein active site with a coenzyme (thiamine pyrophosphate) as the catalytic element of pyruvate decarboxylase

functional bond between a protein and a ligand is usually noncovalent and therefore reversible. Covalent bonds formed in certain catalytic processes are usually short-lived and do not contribute to the stability of the product). In addition to providing a suitable area for reactions, hydrophobic environments also play an important role in determining the orientation of substrates. Any polar moiety in a predominantly hydrophobic area (i.e., the active site) stands out like a beacon and may determine the alignment of the substrate (Fig. 1.27).

Chemical affinity of complexes may be expressed in terms of their dissociation constant $K$, where $k_1$ and $k_2$ indicate the rate of formation and degradation of the complex:

$$K = k_2/k_1$$

The dissociation constant determines the state of equilibrium in antibodies, receptors, transport proteins, and other protein-ligand complexes. For enzymes, the equation is supplemented by a third variable indicating the rate at which the complex converts into its final product ($k_3$):

$$K = (k_2 + k_3)/k_1$$

**Fig. 1.27** The role of background conditions in exposing the object of interest. Candlelight is clearly visible in a dark background (**a**) but becomes hard to discern when the background is well lit (**b**)

In enzymatic reactions $K$ is also called the Michaelis–Menten constant ($K_M$) in honor of Leonor Michaelis and Maud Menten who discovered the relation between reaction rate $v$ and substrate concentration $S$, showing that the equation

$$v = V_{MAX}{}^*[S]/(K_M + [S])$$

determines the so-called saturation curve and that further concentration of the substrate beyond its maximum value for a given concentration of the enzyme does not result in an increased reaction rate. This observation yielded two important parameters of enzymatic reactions ($K_M$ and $V_{MAX}$) and contributed to many important discoveries involving enzymes.

The activity of enzymes goes beyond synthesizing a specific protein-ligand complex (as in the case of antibodies or receptors) and involves an independent catalytic attack on a selected bond within the ligand, precipitating its conversion into the final product. The relative independence of both processes (binding of the ligand in the active site and catalysis) is evidenced by the phenomenon of noncompetitive inhibition, where the Michaelis-Menten constant remains unchanged despite a cessation of enzymatic activity in the presence of an inhibitor. This proves that the process of catalysis (dependent on coordinated action of numerous amino acids and, optionally, a coenzyme) is orthogonal to the enzyme's capability to bind the ligand. Kinetic studies of enzymes have provided valuable insight into the properties of enzymatic inhibitors—an important field of study in medicine and drug research. Some inhibitors, particularly competitive ones (i.e., inhibitors which outcompete substrates for access to the enzyme), are now commonly used as drugs.

While proteins are among the most basic structures whose involvement in biological processes may be described as *active* and in spite of the fact that nearly all biological activity is protein-based, it should be noted that some biological processes occur on the RNA level. Ribozymes (which perform a handful of biological functions) appear to be among the most primordial active structures in biology.

**Table 1.1**   General properties of proteins originating from cells and the organism

| Location | Intracellular proteins | Membrane proteins | Blood proteins | Interstitial fluid proteins |
|---|---|---|---|---|
| Structural properties | *Proteins found in intracellular fluids* Usually globular, devoid of saccharide components *Scaffold proteins* Usually fibrillar | Usually globular, chiefly glycoproteins Moderately or strongly hydrophobic surfaces | Globular, with polar surfaces; mostly glycoproteins | Mostly fibrillar; fused with proteoglycans |
| Functional properties | *Proteins found in intracellular fluids* Mostly enzymes; proteins responsible for motion and regulation *Scaffold proteins* Mechanical stabilization of the cell | Receptor proteins, integrated transport proteins | Transport Protection: – Immune response – Blood coagulation – Inhibitors | Mostly support-related functions |

Proteins are capable of functioning both as receptors and effectors. They may also actively transmit biological signals (e.g., kinases), although the role of passive carriers usually falls to non-protein structures (peptides, steroids, etc.) The distinction between receptor and effector proteins is the basic premise of this handbook. It applies to all functional proteins, irrespective of their location and task. Effector structures include enzymes, ion pumps and channels, proteins involved in transcription of genetic material, proteins which regulate muscle contraction, etc.

With respect to the locations and properties of proteins, we can generally distinguish the following groups (see Table 1.1):

1. Cellular proteins
2. Proteins belonging to the organism level

Since cells and organisms exist as distinguishable (though interrelated) biological entities, their respective proteins serve different purposes and can therefore be treated as separate groups. Cellular proteins are found inside cells and in cellular membranes, while organism's proteins are present in interstitial cavities and in blood.

Cellular functional proteins are usually globular and often associated with enzymatic or regulatory functions, although in certain types of cells (such as myocytes) proteins governing contraction and relaxation may have a fibrillar structure. Another important subgroup consists of cytoskeletal proteins.

Membrane proteins also tend to be globular; however they are clearly distinct from intracellular proteins with respect to their structure and task. Unlike water-soluble proteins, they expose hydrophobic residues on their surface. This property is required for integration with cellular membranes. Membrane proteins responsible for interfacing with the external environment include receptors, pumps, and structures which facilitate cell adhesion and interaction.

Interstitial structures are apart from polysaccharides mostly composed of support proteins. A wholly distinct group of proteins can be found in blood. Their task is to mediate organ-to-organ communication, ensure homeostasis, and perform a wide array of protective functions, e.g., immune response, blood coagulation, inhibition of proteolytic enzymes, etc.

### 1.1.5   Energy and Information Storage Structures

Sequestration of resources is subject to two seemingly contradictory criteria:

1. Maximize storage density.
2. Perform sequestration in such a way as to allow easy access to resources.

In order for any system to gracefully tolerate variations in the availability of raw resources and demand for products, storage capabilities are required. Such capabilities augment the system's autonomy, permitting continued operation even when crucial resources are temporarily lacking or products cannot be disseminated. In most biological systems, storage applies to energy and information. Other types of resources are only occasionally stored (this includes, e.g., iron, which is consumed in large amounts yet infrequently available—sequestration of iron is mediated by a dedicated protein called ferritin).

Energy is stored primarily in the form of saccharides and lipids. Saccharides are derivatives of glucose, rendered insoluble (and thus easy to store) via polymerization. Their polymerized forms, stabilized with $\alpha$-glycosidic bonds, include glycogen (in animals) and starch (in plantlife). An important side effect of this type of bonding is the formation of twisted polymer chains, limiting packing density but enabling easy access. Glycosidic links are usually of type 1–4 (sometimes 1–6), giving rise to branching, treelike structures which can be readily manipulated by enzymes due to the multitude of potential points of contact (see Fig. 1.16).

It should be noted that the somewhat loose packing of polysaccharides, compounded by the presence of partially oxygenated carbons and relatively high degrees of hydration, makes them unsuitable for storing large amounts of energy. In a typical human organism, only ca. 600 kcal of energy is stored in the form of glycogen, while (under normal conditions) more than 100,000 kcal exists as lipids. Lipid deposits usually assume the form of triglycerides (triacylglycerols). Their properties can be traced to the similarities between fatty acids and hydrocarbons. Storage efficiency (i.e., the amount of energy stored per unit of mass) is twice that of polysaccharides, while access remains adequate owing to the relatively large surface area and high volume of lipids in the organism.

Most living organisms store information in the form of tightly packed DNA strands. Once neutralized with histones (in eukaryotic organisms), DNA can be wound on a protein scaffold, as depicted in Fig. 1.28.

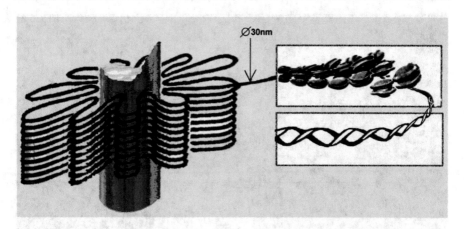

**Fig. 1.28** Hypothetical packing of DNA strands into chromosomes. The figure depicts consecutive stages of DNA coiling. Relative dimensions of elements are distorted in order to emphasize packing strategies

The principles of efficient packing and easy access apply here as well, precluding the simplest possible packing of DNA—multilayered coiling (akin to winding a thread on a spool). Similarly, all forms of solid-like aggregations are inefficient and therefore unusable. The most effective form of DNA packing involves looping the strand and winding it around a chromosomal scaffold so that each loop can be accessed separately and each fragment uncoiled without unwinding the entire chain. This model can roughly be compared to the arrangement of books on library shelves, enabling unobstructed access to each book (Fig. 1.29).

DNA packing begins by arranging it on histones. However, final packing consists of several stages. Histone-centered loops can be further coiled, creating a tight spiral with a microscopic thickness of 30 nm. In such a coiled thread, each loop is independently attached to a protein scaffold which forms the backbone of the chromosome. The final rate of compression (compared to the length of the uncoiled strand) is approximately 10,000-fold. It should be noted that only a small percentage of DNA (about few %) conveys biologically relevant information. The purpose of the remaining ballast is to enable suitable packing and exposure of these important fragments. If all DNA were to consist of useful code, it would be nearly impossible to devise a packing strategy guaranteeing access to all of the stored information.

## 1.2 Self-Organization

All complex cellular structures are the result of self-organization. In this process, simple elements spontaneously generate complex structures by connecting to one another in a predetermined way, usually with noncovalent bonds. The function of the cell is limited to synthesis of startup elements for self-organization. Figure 1.30

**Fig. 1.29** The library as an information storage model, allowing easy access to books. Similar strategies can be observed in DNA packing

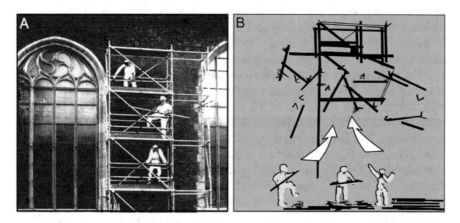

**Fig. 1.30** Organization and self-organization in the creation of coherent structures (schematic depiction). Construction of a scaffold via (**a**) deliberate activity and (**b**) spontaneous process. In the latter case, chaotically arranged elements with appropriate latches spontaneously assemble into a coherent structure

depicts this phenomenon by comparing the concept of self-organization with deliberate ordering.

Controlled self-organization applies only to evolved structures oriented on association, e.g., water-soluble structural components which are thermodynamically unstable (metastable) and try to maximize their relative stability by reaching a global energy minimum. For obvious reasons, this process usually applies to partly polar structures immersed in aqueous solutions; however it may also be observed in fully polar structures where tight packing eliminates the possibility of interaction with water (note that the presence of water tends to inhibit noncovalent interactions between separate units). The cooperative interaction may be the driving force facilitating removal of water, e.g., folding of α-helical polypeptide chains stabilized by hydrogen bonds.

The structure of the aggregate depends on the structure of its components. Self-association yielding coherent structures can be viewed as a type of self-organization. Examples of self-association include the growth of planar phospholipid micelles and folding of polypeptide chains. In general, self-organization is most often based on noncovalent interactions where the associable domain elements achieve (through random collisions) the most energetically stable form for a given set of environmental conditions. Fibrillar supporting structures are intrinsically passive, and their aggregation is a consequence of relatively large contact surfaces. On the opposite end of the activity spectrum, cells guide the process of self-organization by altering the concentrations of substrates and the order in which they are exposed; however, they are usually unable to directly affect the process itself (with certain exceptions where the cell synthesizes structures capable of interfering with self-organization—such as chaperones).

Self-organization may also occur as a result of specific interactions such as the contact between proteins and ligands in active sites.

Fibrillar molecules tend to form parallel clusters. If special structures (e.g., planar or three-dimensional forms) are required or if atypical separation between parallel monomers becomes necessary, the cell may synthesize fibrillar units with attached globular fragments. Such fragments condition the emergence of specific aggregates (a similar process can be observed in the development of intermediate filaments). In most cases, however, the final form of the product is a result of the function of independent globular proteins, sometimes called accessory proteins (Fig. 1.31).

Cellular synthesis of accessory proteins determines, to a large extent, the outcome of self-organization processes. This is similar to the final assembly of furniture or pipe segments using connectors in order to achieve the required configurations (Fig. 1.32).

## 1.3 Hypothesis

### 1.3.1 Protein Folding Simulation Hypothesis: Late-Stage Intermediate—Role of Water

**The In Silico Protein Folding Process**
In biology, function is enabled by proteins. Interestingly, only the primary structure of a protein—its polypeptide sequence—is genetically encoded; however, function can only be expressed when the resulting polypeptide chain adopts a specific 3D

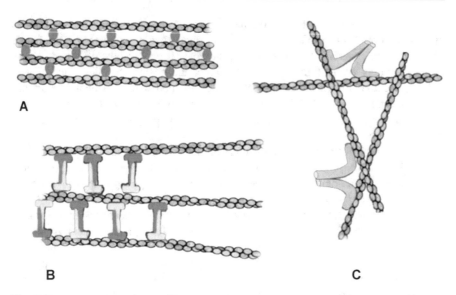

A

B                                            C

**Fig. 1.31** Arrangement of microfilaments enforced by accessory proteins: (**a**) fimbrin, (**b**) α-actinin, and (**c**) filamin

**Fig. 1.32** The role of accessory elements in construction tasks

conformation through folding. Protein folding is a spontaneous process, guided mainly by changes in entropy: $\Delta G = \Delta H - T\Delta S$. It is mediated by interactions between amino acid residues (of varying polarity) and the environment, which is usually represented by water (highly polar) or the surface of a cellular membrane (nonpolar), which anchors some proteins. In addition, folding may be affected by local contact with other proteins or non-protein structures, such as polysaccharides. Finally, the structure of the protein may depend on the sequence of events accompanying its synthesis process. Secondary structure may begin emerging even before the chain has been fully synthesized and released by the ribosome, resulting in a different conformation than ab initio folding of a fully synthesized chain. The presence of a polar solvent promotes the formation of a hydrophobic core, which consists of hydrophobic residues encapsulated by a polar surface. However, the primary sequence does not always enable such arrangement of residues in the 3D protein body. In native proteins, certain deviations from the "ideal" distribution (which consists of a centralized hydrophobic core and a uniform hydrophilic shell) are evolutionarily conditioned and directly related to the protein's function: they correspond to its active sites.

Proteins which share functional characteristics, such as immunoglobulins, albumin, or hemoglobin, usually go through similar folding stages. This enables us to predict their conformation based on the crystal structure of a sample polypeptide. Such techniques tell us *how* proteins fold, but do not provide knowledge of the underlying process. In contrast, here we try to answer a different question: *why* do proteins adopt certain conformations?

## 1. *Protein Folding*

The "oil drop" model is a long-standing abstraction of protein structure, originally proposed by W. Kauzmann [lit]. It treats the protein molecule as a two-layer construct consisting of a polar surface layer and a strongly hydrophobic center (the "core"). According to this model, tertiary structure is stabilized—in addition to disulfide bonds—by a specific distribution of hydrophobicity. The model we propose, referred to as the fuzzy oil drop model, modifies this discrete binary structure by introducing a continuous hydrophobicity gradient, mathematically expressed as a 3D Gaussian superimposed onto the target molecule. This gradient peaks at the center of the ellipsoid and decreases along with distance from the center, reaching near-zero values at a distance of $3\sigma$ (this is known as the three-sigma rule). The corresponding abstraction reflects structural similarities between bipolar molecules, i.e., substances, which form micellar structures in an aqueous solvent (such as soap). Notably, amino acids are also bipolar (with varying levels of polarity), and the presence of water plays a major role in the folding process. The polar solvent promotes exposure of polar residues on the surface and directs hydrophobic residues toward the center of the emerging structure, ensuring entropically advantageous conditions at the protein/water interface. Note, however, that the protein usually cannot form a perfect "micelle-like" structure due to the presence of covalent bonds between amino acids, restricting their mobility—this is why perfect micellar order is rarely observed in native proteins (note also that this would render the protein incapable of interacting with any molecules other than water or solvated ions). Some classes of proteins produce near-perfect micellar structures; these include downhill, fast-folding, ultra-fast-folding, or type III antifreeze proteins—all indicative that the presence of water is a crucial aspect of the folding process and is also sufficient for the protein to develop a hydrophobic core along with a hydrophilic shell. Figure 1.33 illustrates proteins in which the distribution of hydrophobicity is consistent with the Gaussian form (linear and 2D presentation).

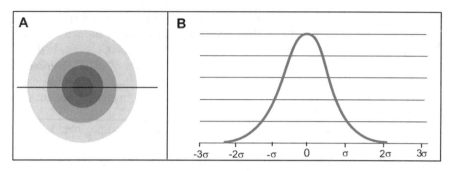

**Fig. 1.33** Idealized distribution of hydrophobicity consistent with the Gaussian distribution: (**a**) 2D presentation (shades of gray correspond to increasing levels of hydrophobicity) and (**b**) linear presentation (simple Gaussian)

### How Function Is Encoded as Distribution of Hydrophobicity

The presence of an idealized micelle-like distribution (matching a 3D Gaussian form)—as remarked above—promotes high solubility but also prevents the protein from interacting with other specific molecules, which is a condition of biological activity. This is why, in addition to fragments which exhibit a micellar order, in vivo proteins also include discordant fragments where the observed distribution of hydrophobicity deviates from the theoretical pattern. This situation can be quantitatively modeled using Kullback-Leibler's divergence entropy model, which quantifies the contribution of accordant and discordant fragments to the protein's overall distribution of hydrophobicity (consisting of a hydrophobic core and a polar shell).

Divergence entropy, sometimes also referred to as distance entropy, expresses the difference between the target distribution (in our case—the 3D Gaussian) and the distribution actually observed in the given structure (which, in proteins, depends on hydrophobic interactions between amino acid residues).

Figure 1.34 illustrates a discordant distribution of hydrophobicity, where strongly hydrophobic residues are not concentrated at the center of the protein body.

### 2. *Local Excess of Hydrophobicity on the Protein Surface*

Another deviation from the theoretical model, known to be associated with biological activity, involves exposure of hydrophobicity on the protein surface. Proteins which follow this structural pattern tend to associate with other molecules, including other proteins, forming quaternary structures (Fig. 1.35) which can be regarded as containing a single, shared hydrophobic core. Exposed hydrophobicity may therefore indicate a potential complexation site, capable of attracting a partner protein which also exposes hydrophobic residues on its own surface.

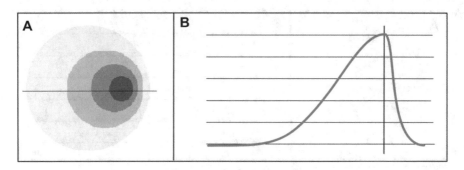

**Fig. 1.34** Example distribution with a displaced hydrophobic core. (**a**) 2D presentation (shades of gray correspond to increasing levels of hydrophobicity) and (**b**) linear presentation (skewed Gaussian)

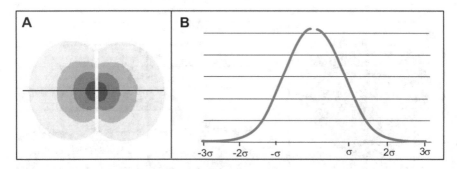

**Fig. 1.35** Discordant distribution of hydrophobicity—exposure of hydrophobicity on the protein surface. (**a**) 2D presentation (shades of gray correspond to increasing levels of hydrophobicity); the black line corresponds to the cross-section depicted on the right-hand side. (**b**) Linear presentation (two distorted Gaussians, which together form a centralized hydrophobic core for the dimeric structure)

### 3. *Local Deficit of Hydrophobicity*

Another type of discordance involves a local deficit of hydrophobicity, typically associated with the presence of a binding cavity (Fig. 1.36).

Hydrophobicity distribution analysis (Figs. 1.34 and 1.35) reveals that a large portion of the protein body follows a distribution which approximates the 3D Gaussian, thus ensuring that the protein remains soluble (i.e., exposes a polar surface to its environment).

### 4. *Effect of Non-aqueous Environments on the Folding Process*

An entirely different situation is observed in the case of membrane proteins, surrounded by a strongly hydrophobic layer of hydrocarbons, which is part of the cell membrane. In order to achieve stability, a protein anchored in this manner (e.g., rhodopsin) must expose hydrophobic residues on its surface. Additionally, if the

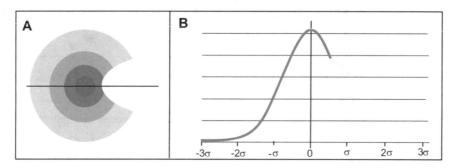

**Fig. 1.36**  Distribution of hydrophobicity showing a local deficit compared with the theoretical 3D Gaussian. (**a**) 2D presentation (shades of gray correspond to increasing levels of hydrophobicity); the black line corresponds to the cross-section depicted on the right-hand side. (**b**) Linear presentation (truncated Gaussian associated with the presence of a cavity)

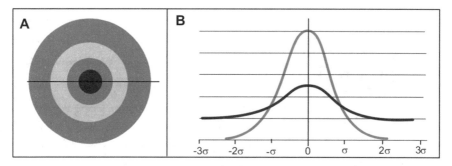

**Fig. 1.37**  Hydrophobicity distribution in a membrane protein. (**a**) 2D presentation, showing excess hydrophobicity on the surface (where the protein contacts the membrane). (**b**) Linear presentation, illustrating differences between the observed (black) and theoretical (blue) distribution

protein is meant to interact with a ligand, its central part must include a suitable polar cavity. Such deficit of hydrophobicity in the protein's interior is particularly noteworthy in light of the presented model, which suggests a concentration of hydrophobicity in the core. Figure 1.37 shows a sample system with these properties.

Taken together, these properties result in a distribution of hydrophobicity which deviates from the 3D Gaussian and may even be its polar opposite. An example is provided by ion transport proteins, which expose hydrophobic residues in order to anchor themselves in the cell membrane and also contain polar cores, with a central cavity surrounded by hydrophilic residues. This is schematically depicted in Fig. 1.38.

### 5.  Complex Structure of Ion Transport Proteins

Such proteins are composed of multiple chains, each of which includes a transmembrane domain as well as a terminal domain (or domains) exposed to the aqueous

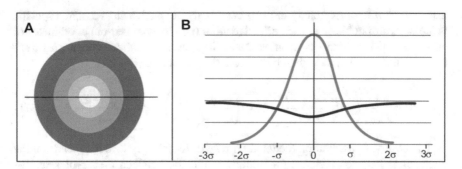

**Fig. 1.38** Distribution of hydrophobicity in an ion transport protein. (**a**) 2D presentation (shades of gray correspond to increasing levels of hydrophobicity). (**b**) Linear presentation, revealing differences between a centralized hydrophobic core (blue) and the observed distribution in proteins which act as ion transport channels (black)

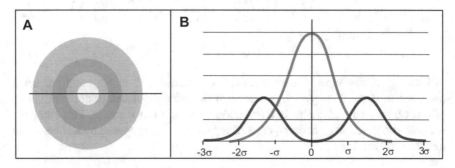

**Fig. 1.39** Idealized distribution of hydrophobicity in the terminal domain of an ion transport protein. (**a**) 2D presentation (shades of gray correspond to increasing levels of hydrophobicity)—the black horizontal line corresponds to the cross-section illustrated on the right-hand side. (**b**) Linear presentation, revealing differences between the idealized 3D Gaussian (blue) and the modified distribution (black)

environment. In the latter case, the central part is hydrophobically deficient (due to the presence of a channel and its polar surroundings, enabling contact with the solvent). Figure 1.39 illustrates an idealized distribution of hydrophobicity in a terminal domain.

### 6. *Hydrophobicity Distribution Parameters*

So far we've focused on visual interpretations of the presence of the aqueous solvent and its effect on the folding process. In mathematical terms, the presented model can be described as follows.

As already suggested, W. Kauzmann's original oil drop model has been extended into what we refer to as the fuzzy oil drop (FOD) model. In FOD, the discrete distribution of hydrophobicity gives way to a continuous distribution, mathematically expressed as a 3D Gaussian form. This Gaussian corresponds to the idealized distribution of hydrophobicity in the protein body. It enables us to assign a specific value of hydrophobicity to each effective atom (i.e., to averaged-out coordinates of all atoms which comprise a given residue). The resulting idealized micellar distribution is denoted T and described by the following function (Eq. 1.1):

$$H_i^T = \frac{1}{H_{\mathrm{sum}}^T} \exp\left(\frac{-(x_i - \bar{x})^2}{2\sigma_x^2}\right) \exp\left(\frac{-(y_i - \bar{y})^2}{2\sigma_y^2}\right) \exp\left(\frac{-(z_i - \bar{z})^2}{2\sigma_z^2}\right) \quad (1.1)$$

The observed distribution (denoted $O$) may differ from T since it depends on the interactions between residues (or, more specifically, between their representative effective atoms). Such interactions depend, among others, on the separation between residues as well as on their intrinsic hydrophobicity. Specific values of $O_i$ are given by M. Levitt's formula [lit] (Eq. 1.2):

$$H_i^o = \frac{1}{H_{\mathrm{sum}}^O} \sum_j \begin{cases} \left(H_i^r + H_j^r\right)\left(1 - \frac{1}{2}\left(7\left(\frac{r_{ij}}{c}\right)^2 - 9\left(\frac{r_{ij}}{c}\right)^4 + 5\left(\frac{r_{ij}}{c}\right)^6 - \left(\frac{r_{ij}}{c}\right)^8\right)\right), & \text{for } r_{ij} \leq c \\ 0, & \text{for } r_{ij} > c \end{cases}$$

$$(1.2)$$

Here, $r_{ij}$ is the separation between effective atoms, C is the cutoff distance (assumed to equal 9 Å), and $H^r$ is the intrinsic hydrophobicity. Both distributions ($T$ and $O$) may be quantitatively compared following normalization. This comparison relies on the concept of divergence entropy, originally proposed by Kullback and Leibler [lit] (Eq. 1.3):

$$D_{KL}(P|Q) = \sum_{i=1}^{N} P_i \log_2 \frac{P_i}{Q_i} \quad (1.3)$$

In the presented model, the distribution subject to analysis ($P_i$) corresponds to $O_i$, while the reference distribution ($Q_i$) is supplied by $T_i$.

Given the nature of the analyzed quantity (entropy), the need for another reference distribution emerges. This distribution, denoted R, is regarded as opposite to $T$, as it ascribes the same value to each residue in the chain ($R_i = 1/N$, N being the total number of residues). Under such conditions, hydrophobicity is uniformly distributed throughout the protein body, and no hydrophobic core is present. As a result, the status of a given protein may be described using a pair of $D_{KL}$ values:

$$D_{KL}(O|T) = \sum_{i=1}^{N} O_i \log_2(O_i/T_i) \quad (1.4)$$

for the $O|T$ relationship, and

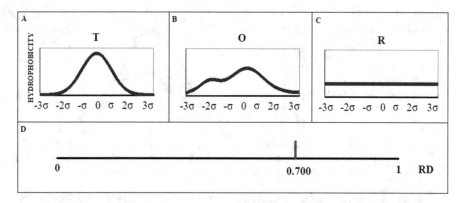

**Fig. 1.40** Visualization of distributions used to determine the status of the given protein. (a) Idealized Gaussian distribution (*T*)—reduced to a single dimension for clarity. (b) Observed distribution (*O*), conditioned by mutual interactions between residues (i.e., between their effective atoms). (c) Linear distribution (*R*) where each residue is assigned the same value of hydrophobicity. (d) RD scale (where 0 corresponds to *T*, while 1 corresponds to *R*). The red line indicates the value of RD calculated for the observed distribution (see item B above); in this case RD = 0.700

$$D_{\mathrm{KL}}(O|R) = \sum_{i=1}^{N} O_i \log_2(O_i/R_i) \tag{1.5}$$

for the *O*|*R* relationship.

In some respects, the *R* distribution may be regarded as analogous to "vacuum conditions," emerging in the absence of any external factors which might influence the distribution of hydrophobicity in a protein chain.

The overall status of the protein depends on the relation between DKL values computed for the *O*|*T* and *O*|*R* configurations, respectively. If the former value is lower, the protein is said to possess a centralized hydrophobic core. This is schematically depicted in Fig. 1.40.

In order to avoid having to deal with two separate coefficients for a single object, we introduce the RD (relative distance) parameter, which is expressed as follows:

$$\mathrm{RD} = \frac{D_{\mathrm{KL}}(O|T)}{D_{\mathrm{KL}}(O|T) + D_{\mathrm{KL}}(O|R)} \tag{1.6}$$

When RD < 0.5, the protein is assumed to contain a well-defined hydrophobic core and a polar shell—unlike the structure illustrated in Fig. 1.40.

This abstraction enables us to quantitatively assess the similarities and differences between different distributions—whether local (Figs. 1.34, 1.35, and 1.36) or global (Figs. 1.37, 1.38, and 1.39).

7. *Involvement of Environmental Hydrophobic Factors in Shaping the Protein's Structure*

**Fig. 1.41** Distribution shown in Fig. 1.40 superimposed onto a profile which satisfies $K = 0.5$ (teal curve). The T and O distributions are plotted in blue and brown, respectively

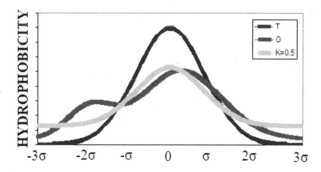

As noted above, global mismatch between the observed distribution of hydrophobicity and the idealized Gaussian is often due to active involvement of the environment, which guides the folding process depending on its own properties. In such cases, the environment contains a hydrophobic factor, which can be recognized in the fuzzy oil drop model as a function complementary to the original Gaussian, denoted $M_i$ (where 3DG corresponds to the $T$ distribution):

$$M_i = [1\text{–}3\text{DG}_i]_n \tag{1.7}$$

It turns out, however, that even membrane proteins do not follow the distribution described by Eq. (1.7). The presence of water is ubiquitous, and therefore—in addition to any hydrophobic factors—the aqueous solvent must be acknowledged in our calculations. Thus, our description of the external force field is extended as follows:

$$M_i = \left[3\text{DG}_i + (1\text{–}3\text{DG}_i)_n\right]_n \tag{1.8}$$

In practice, the $M$ distribution is calculated by substituting $T_{\text{MAX}}$ for "1." Furthermore, we introduce the $K$ parameter, which reflects the varying proportions of the aqueous and hydrophobic environments. This yields the final equation:

$$M_i = \left[3\text{DG}_i + \mathbf{K}(T_{\text{MAX}}\text{–}T_i)_n\right]_n \tag{1.9}$$

(The $n$ index corresponds to normalization of the distribution.)

Analysis of various proteins reported in literature indicates that $K = 0$ is observed in structures listed at the beginning of this subsection, downhill, fast-folding, ultra-fast-folding, and type III antifreeze proteins—all of which contain well-defined hydrophobic cores, encapsulated by polar shells. This condition is also fulfilled by the nearly all protein domains (when analyzed as standalone units).

For the vast majority of proteins, particularly single-chain proteins, $0 < K < 0.5$. This group also includes some enzymes and small proteins which have a quaternary structure (mainly homodimers), as shown in Figs. 1.34, 1.35, and 1.36.

Figure 1.41 illustrates the role of the $K$ coefficient. The idealized distribution ($T$), where the aqueous solvent directs all hydrophobic residues toward the center of the protein body while exposing polar residues on its surface, is not a good match for the observed distribution ($O$). Modifying $T$ with $K = 0.5$ results in a target distribution which more closely reflects the observed values (brown curve). This suggests that the protein in question may have folded in an environment which, in addition to water, includes a so-called chaotropic factor [lit], whose involvement is mathematically expressed by the value of $K = 0.5$.

For membrane proteins, $K$ may adopt values greater than 0.5 or even greater than 1.0. Examples of such proteins (where, in some cases, $K > 3$) include bacterial efflux pumps.

Analysis of a large group of proteins using RD and $K$ parameters shows that the fuzzy oil drop model, in its modified form (FOD-M), is a useful tool in the study of protein structures. It also suggests that protein structure prediction tools which rely on numerical algorithms should acknowledge the variability of the external environment along the lines proposed above. In each case, the environment contributes a force field which plays an active role in shaping the structure of the emerging protein and affects its biological properties.

The notion that "sequence determines the 3D structure of a protein" reflects the structural encoding of functional properties. In fact, we can more accurately state that "sequence determines the scope and degree of the inability to generate a perfect micellar structure." The values of RD and $K$ determine the specific properties of the given conformation, which may only be capable of biological activity in the presence of external factors other than water—for example, the cellular membrane.

**Misfolding**
Our discussion of protein structure analysis should also acknowledge the pathological phenomenon known as misfolding. This is something that occurs in living organisms on an ongoing basis; however, organisms have evolved methods to eliminate incorrectly folded proteins—for example, through the so-called unfolded protein response (UPR) mechanism [lit].

Misfolding is particularly notable as the causative factor promoting formation of amyloid plaque. In general, globular proteins tend to follow the Gaussian distribution (Fig. 1.42a), whereas amyloid fibrils consist of unit chains, each of which is planar rather than globular. It is, in fact, possible to assess the presence of a hydrophobic core in each of these planar units—however, such cores are two-dimensional and therefore modeled by a 2D Gaussian (Fig. 1.42b). As a result, the unit chain only exposes polarity at the edges of its "disc," whereas both sides contain unshielded hydrophobic residues (Fig. 1.42c), which, in turn, promotes complexation with other similarly shaped chains. This is why amyloid fibrils tend to exhibit unbounded growth.

The description of the external force field proposed in FOD and FOD-M models can be applied to any type of environment, as well as to any input chain with specific chemical properties.

Based on the FOD-M model, we may propose an amyloid transformation mechanism, which turns out to be variable in scope. The transformation of the native protein into its amyloid form can be accompanied by a decrease or an increase in the

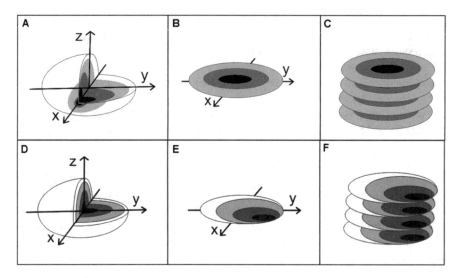

**Fig. 1.42** Distribution of hydrophobicity (denotes by shades of gray) in (**a**) globular protein roughly consistent with the 3D Gaussian (high $K$). (**b**) Individual unit chain of an amyloid fibril, consistent with the 2D Gaussian (low $K$). (**c**) Larger fragment of an amyloid fibril, where the clustering of individual unit chains produces a distribution of hydrophobicity roughly consistent with the 3D Gaussian (low $K$). (**d**) Regularly shaped protein following redistribution of hydrophobicity (low $K$). (**e**) Transformation results in a distribution which differs from theoretical values (high $K$). (**f**) Ultimately, the entire fibril deviates from the 3D Gaussian distribution (high $K$)—this is assumed to result from the presence of chaotropic agents

value of $K$. The specific mode of transformation may be assessed by comparing WT structures and their amyloid counterparts. Two distinct modes can be distinguished:

Amyloid transformation accompanied by a decrease in the value of $K$ occurs, e.g., in the case of alpha-synuclein. The native protein deviates significantly from a globule, and its value of $K$ is very high. Its structure has been determined in complex with a micelle mimicking the axon terminals of presynaptic neurons with which alpha-synuclein associates in vivo. The amyloid structure listed in Protein Data Bank, however, shows a low value of $K$ and reveals the presence of a hydrophobic core. This scenario can be interpreted as follows:

The high value of $K$ results from the presence of a target, which alters the properties of the surrounding environment and forces the protein to adopt a non-globular conformation. Once this external factor is removed, alpha-synuclein can produce a micellar structure, which comprises a centralized hydrophobic core along with a hydrophilic shell—the latter formed by loose, unstructured fragments of the chain (residues 1–30 and 100–140). Therefore, the amyloid structure of alpha-synuclein emerges as a result of interaction with the aqueous solvent in conditions where free protein molecules (detached from the scaffold) are abundant (see Fig. 1.43a and b).

The opposite process—i.e., amyloid transformation accompanied by a major increase in $K$—is observed for the $L$ domain of the IgG light chain. Here, too, the

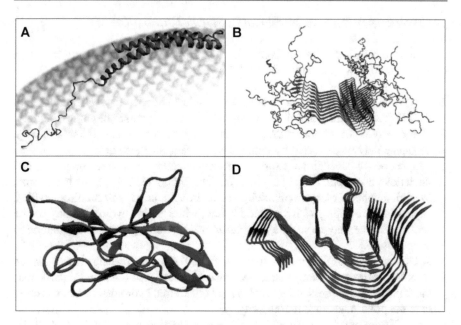

**Fig. 1.43** Variable modes of amyloid transformation: (**a**) Alpha-synuclein in its native form, in complex with axon terminals of presynaptic neurons (experimentally simulated by a micelle)—high value of $K$. (**b**) Alpha-synuclein in its amyloid form ($K < 0.5$, which represents dominant influence of the aqueous environment). (**c**) VL domain of immunoglobulin $G$—low value of $K$; globular structure. (**d**) VL domain of immunoglobulin $G$ in its amyloid conformation—high value of $K$ caused by external factors

external environment plays a crucial role, although in this particular case amyloid transformation is caused by introduction of nonstandard external factors (Fig. 1.43c and d).

We can conclude that in both cases the environment is an important driver of amyloid transformation. With regard to alpha-synuclein, it acts as a scaffold, maintaining a high value of $K$ and ensuring proper biological activity of the protein. In other cases high $K$ may be caused by the unusual external factors—for example, shaking, which is known to promote amyloid transformation in experimental studies. Shaking causes aeration of the solvent and greatly increases the surface area of the water-air interface. Under these conditions water loses its ability to properly guide the folding process of solvated polypeptide chains. On the other hand, in the case of alpha-synuclein, water appears to cause the protein to adopt a non-native conformation which is typical for globular proteins. This is schematically depicted in Fig. 1.43.

Fragments marked in red participate in formation of the amyloid fibril.

1. *Universality of the FOD Model and Its Modified Form (FOD-M)*

When summarizing the outcomes of applying the modified fuzzy oil drop (FOD-M) model, we can conclude the following:

1. The protein is effectively an "intelligent micelle," in which local deviations from the 3D Gaussian distribution of hydrophobicity correspond to biological activity.
2. The widely repeated notion that the protein's 3D structure is determined by the sequence may be extended by noting that the sequence also determines the type and scope of deviations from a perfect micellar structure. Such targeted deviations correspond to sites which mediate biological activity—for example, local deficiencies of hydrophobicity are often associated with binding cavities, capable of interacting with a specific ligand, while excess hydrophobicity on the protein surface may indicate a complexation site which gives rise to quaternary structures.
3. Applying the FOD-M model to fully folded proteins reveals factors which stabilize the protein's native structure, ensuring biological function. In particular, the $K$ parameter expresses the involvement of external hydrophobic factors (other than the polar aqueous environment).
4. The $K$ parameter is also useful in folding simulations—it would, after all, be impossible to explain the extreme structural diversity of in vivo proteins with a mechanism which does not acknowledge the influence of the external environment. The biological properties of a protein must reflect the properties of the environment in which that protein is expected to express its function.
5. In addition to characterizing the protein's structure (discordance vs. a globular conformation which includes a centralized hydrophobic core), the $K$ parameter also conveys the involvement of external "chaotropic" factors which disrupt the natural structure of the aqueous solvent. High values of $K$ indicate that water no longer plays a key role in guiding local processes.
6. Another phenomenon worth taking into account is the structure of amyloid fibrils, dominated by inter-chain hydrogen bonds. In such systems, each residue enters into two hydrogen bonds with adjacent chains, and this process is repeated across nearly the entire chain. The presence of a "chaotropic" factor likely promotes formation of hydrogen bonds.

**Amyloid Transformation**

Under certain conditions, the globular structure of a soluble protein may yield an entirely different distribution of hydrophobicity, leading to planar conformations, mathematically described by a 2D Gaussian. The properties of the underlying mechanism remain an open question; it can, however, be assumed, that the globular structure emerges as a result of a favorable change in $\Delta S$, guided by the polar environment which promotes exposure of hydrophobicity on the surface and concealment of hydrophobic residues within the core. On the other hand, amyloid fibrils are primarily stabilized by hydrogen bonds, which involve all N–H and C=O groups of a given peptide. Such significant involvement of hydrogen bonds suggests that the process is instead driven by $\Delta H$. The causes of such a change remain unclear; however, in later chapters we propose some mechanisms which might explain the amyloid transformation process.

As already mentioned, all known amyloid structures consist of planar unit chains. In this sense, amyloid transformation may be compared with migrating from a 3D to a 2D Gaussian. This is particularly evident in cases where the central part of the cross-section (perpendicular to the fibril's axis) is occupied by hydrophobic residues.

Conformational changes which accompany amyloid transformation require further analysis; however, amyloids are generally dominated by beta folds, separated by short loops. Hydrogen bonds—a characteristic feature of beta structures—link atoms belonging to adjacent chains. In this case, the conformational changes occurring in helical segments appear obvious; however, it is more difficult to explain the transformation between beta folds in native proteins and different beta folds in their amyloid counterparts.

Referring again to the 3D and 2D Gaussian forms, it should be noted that the helix is clearly an example of a 3D structure, given the presence of side chains which radiate outward from the central "hub" in a spiral fashion. Planar structures may therefore only emerge on the basis of beta folds.

Amyloid transformation may also be studied by comparing the differences between dihedral angles (Phi and Psi) in the native form and its amyloid counterpart. While in helices and native beta folds the distribution of these angles is relatively broad, amyloid structures are limited to a very narrow range of angles—for instance, in the case of transthyretin (native structure, 1DVQ; amyloid form, 6SDZ), as evidenced by Ramachandran plots (Fig. 1.44).

In helical fragments, the values of angles are always limited to very narrow range which roughly corresponds to an "idealized" helix. On the other hand, when looking at beta folds, we can see that in amyloids these angles occupy a much smaller range than in native proteins. It is also worth noting that the data points are distributed along a specific line (the diagonal).

When confronting the distribution of Phi and Psi angles with geometric parameters (curvature radius and twist angle between planes of adjacent peptide bonds), we can notice that values of $V = 0$ and $V$-180 are preferred. Hydrogen bonds tend to produce parallel structures, without any twist, as seen in all beta and beta-barrel structures (Fig. 1.44a). This perfectly parallel arrangement of hydrogen bonds is only possible when V becomes equal to 0 or 180° (Fig. 1.44b). In contrast, the distribution of Phi and Psi angles suggests that in native proteins the curvature radii may vary even when it comes too beta sheets. This dispersion of curvature radii produces tertiary structural artifacts, as seen, e.g., in wild-type transthyretin, where $V$ values which deviate from 0 or 180 give rise to an arched form, seen in proteins which contain secondary or supersecondary beta fragments.

Structural changes are evident when comparing the three fragments. In the native form, the beta sheet is twisted due to changes in orientation of adjacent peptide bond planes (where $V$ is close to, but not exactly equal to, 180°—as seen in Fig. 1.44c). In the amyloid structure, $V$ becomes equal to 180. One of the fragments becomes a

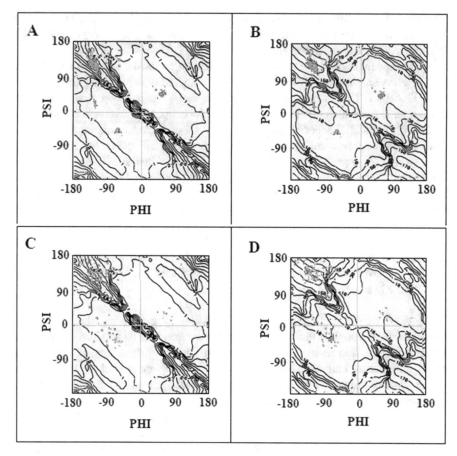

**Fig. 1.44** Distribution of Phi and Psi angles (blue dots) on Ramachandran plots in transthyretin. (**a** and **b**) Amyloid structure and (**c** and **d**) native structure. (**a** and **c**) Ramachandran plot indicating zones which correspond to curvature radii for a given pair of Phi and Psi angles. (**b** and **d**) Ramachandran plot indicating zones which correspond to twist angles between adjacent peptide bond planes in a polypeptide chain. Calculation of these parameters is also described in Sect. 3.7.1. See also Protein Folding In Silico—Woodhead Publishing currently Elsevier 2012, Ed: Irena Roterman-Konieczna Oxford, Cambridge, Philadelphia, New Dehli 2012

loop, but the remaining two run parallel to each other, enabling the formation of hydrogen bonds. It is also notable that a beta sheet consists of fragments contributed by the same polypeptide chain, whereas in an amyloid there is contact between beta folds belonging to independent chains. This process involves transformation from a 3D to a 2D structure.

Our analysis may also be applied to loops responsible for creating the 3D structure. It should be noted that the set of points on the Ramachandran plot which represents the native structure (Fig. 1.45a and b) corresponds to the helical fragment, whereas in its amyloid counterpart, the same points belong to loops which do not adopt a helical conformation.

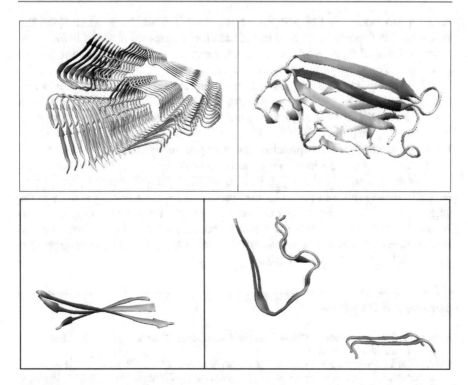

**Fig. 1.45** Structure of individual fragments, highlighting the beta sheet. The fragment which does not retain beta conformation in the amyloid is marked in red. (**a**) Amyloid (6SDZ) and (**b**) native protein (1DVQ)

The range of values plotted for the amyloid conformation of transthyretin is restricted to the maximum curvature radius—i.e., linear structures, where $V = 180°$. This results in a planar arrangement of unit chains, which promotes formation of hydrogen bonds between parallel polypeptides.

Comparing the native and amyloid structures of transthyretin further highlights the conformational changes which accompany amyloid transformation. Here, the relevant question is: what factor contributes to parallel alignment of adjacent polypeptide planes ($V = 0$ or $180°$), enabling hydrogen bonds to form? Some clues might be provided by studying the structural properties of water in its natural state, in physiological saline solutions, as well as in the presence of distorting factors. The presented hypothesis assumes that such factors promote formation of 2D structures while disrupting 3D structures. In particular, it might be interesting to look at the relation between the structural properties of the solvent and the planar and highly symmetrical nature of the hydrogen bond network. Tracing the progressive changes in Phi and Psi angles, as well as the resulting geometric rearrangements, may shed new light on the amyloid transformation process.

The role of environment shall be linked with the issue of information. The information carried by amino acids is not sufficient to cover the information amount necessary to point the Phi and Psi angles for 3D structure determination. The information deficiency is covered by information the source of which is the specificity of external force field which is delivered by the surrounding for folding process. The detailed presentation of hypothesis of amyloid transformation can be found in: From globular proteins to amyloids. Ed: Irena Roterman-Konieczna, Elsevier, Amsterdam, Netherlands, Oxford OX5 3GB UK, Cambridge MA 02139, USA 2020 available on line [https://www.google.pl/books/edition/From_Globular_ Proteins_to_Amyloids/7K9uEAAAQBAJ?hl=pl&gbpv=1&dq=Roterman-Konieczna+From+globular+proteins+elsevier&printsec=frontcover].

The essential progress in studies concerning protein structure largely contributes to the improvement of drug design techniques including techniques of drug distribution in the organism in particular these allowing focusing of drug activity in the target as, for example, technique represented by immunotargeting In: Self-Assembled Molecules—New Kind of Protein Ligands Ed: Irena Roterman and Leszek Konieczny; Springer Open 20188. Available on-line [https://link.springer. com/book/10.1007/978-3-319-65639-7].

## Suggested Reading

Arendt D (2018) Hox genes and body segmentation. Science. 361(6409):1310–1311. https://doi. org/10.1126/science.aav0692

Bai L, You Q, Feng X, Kovach A, Li H (2020) Structure of the ER membrane complex, a transmembrane-domain insertase. Nature 584(7821):475–478. https://doi.org/10.1038/s41586-020-2389-3

Brinkerhoff H, Kang ASW, Liu J, Aksimentiev A, Dekker C (2021) Multiple rereads of single proteins at single-amino acid resolution using nanopores. Science 374(6574):1509–1513. https://doi.org/10.1126/science.abl4381

Buss F, Luzio JP, Kendrick-Jones J (2002) Myosin VI, an actin motor for membrane traffic and cell migration. Traffic 3:851–858

Cao J, Belousoff MJ, Liang Y-L, Johnson RM, Josephs TM, Fletcher MF, Christopoulos A, Hay DL, Danev R, Wootten D, Sexton PM (2022) A structural basis for amylin receptor phenotype. Science 375(6587):eabm9609. https://doi.org/10.1126/science.abm9609

Cao L, Coventry B, Goreshnik I, Huang B, Sheffler W, Joon Sung Park JS, Jude KM, Marković I, Kadam RU, Verschueren KHG, Verstraete K, Russel-Walsh ST, Bennett N, Phal A, Yang A, Kozodoy L, DeWitt M, Picton L, Miller L, Strauch E-M, DeBouver ND, Pires A, Bera AK, Halabiya S, Hammerson B, Yang W, Bernard S, Stewart L, Wilson IA, Hannele Ruohola-Baker H, Schlessinger J, Lee S, Savvides SN, Garcia KC, Baker D (2022) Design of protein-binding proteins from the target structure alone. Nature 605(7910):551–560. https://doi.org/10. 1038/s41586-022-04654-9

Carter NJ, Cross RA (2005) Mechanics of the kinesin step. Nature 435:308–312

Chan CJ, Costanzo M, Ruiz-Herrero T, Mönke G, Petrie RJ, Bergert M, Diz-Muñoz A, Mahadevan L, Hiiragi T (2019) Hydraulic control of mammalian embryo size and cell fate. Nature. 571(7763):112–116. https://doi.org/10.1038/s41586-019-1309-x

Cho NH, Cheveralls KC, Brunner AD, Kim K, Michaelis AC, Raghavan P, Kobayashi H, Savy L, Li JY, Canaj H, Kim JYS, Stewart EM, Gnann C, McCarthy F, Cabrera JP, Brunetti RM, Chhun BB, Dingle G, Hein MY, Huang B, Mehta SB, Weissman JS, Gómez-Sjöberg R, Itzhak DN, Royer LA, Mann M, Leonetti MD (2022) OpenCell: endogenous tagging for the cartography of human cellular organization. Science. 375(6585):eabi6983. https://doi.org/10.1126/science. abi6983

Courbet A, Hansen J, Hsia Y, Bethel N, Park Y-J, Xu C, Moyer A, Boyken SE, Ueda G, Nattermann U, Nagarajan D, Silva D-A, Sheffler W, Quispe J, Nord A, King N, Bradley P, Veesler D, Kollman J, Baker D (2022) Computational design of mechanically coupled axle-rotor protein assemblies. Science 376(6591):383–390. https://doi.org/10.1126/science.abm1183

Cranford SW, Tarakanova A, Pugno NM, Buehler MJ (2012) Nonlinear material behavior of spider silk yields robust webs. Nature 428:72–76

DeWitt MA, Chang AY, Combs PA, Yildiz A (2012) Cytoplasmic dynein moves through uncoordinated stepping of the AAA+ ring domains. Science 335:221–225

Dong D, Zheng L, Lin J, Zhang B, Zhu Y, Li N, Xie S, Wang Y, Gao N, Huang Z (2019) Structural basis of assembly of the human T cell receptor-CD3 complex. Nature. 573(7775):546–552. https://doi.org/10.1038/s41586-019-1537-0

Drewes G, Ebneth A, Mandelkow E-M (1998) MAPs, MAPKs and microtubule dynamics. TIBS 23:307–311

Endres D, Miyahara M, Moisant P, Zlotnick A (2005) A reaction landscape identifies the intermediates critical for self-assembly of virus capsids and other polyhedral structures. Prot Sci. 14:1518–1525

Fowler VM, Vale R (1996) Cytoskeleton. Curr Opin Cell Biol 8:1–3

Greber UF (2016) Virus and host mechanics support membrane penetration and cell entry. J Virol 90(8):3802–3805. https://doi.org/10.1128/JVI.02568-15

Gullberg D (2003) The molecules that make muscle. Nature 424:138–140

Gunasekaran K, Ma B, Nussinov R (2004) Is allostery an intrinsic property of all dynamic proteins? Proteins Struct Func Bioinfo 57:433–443

Hadzipasic A, Wilson C, Nguyen V, Kern N, Kim C, Pitsawong W, Villali J, Zheng Y, Kern D (2020) Ancient origins of allosteric activation in a Ser-Thr kinase. Science 367(6480):912–917. https://doi.org/10.1126/science.aay9959

He S, Del Viso F, Chen CY, Ikmi A, Kroesen AE, Gibson MC (2018) An axial Hox code controls tissue segmentation and body patterning in Nematostella vectensis. Science. 361(6409):1377–1380. https://doi.org/10.1126/science.aar8384

He S, Wu Z, Tian Y, Yu Z, Yu J, Wang X, Li J, Liu B, Xu Y (2020) Structure of nucleosome-bound human BAF complex. Science 367(6480):875–881. https://doi.org/10.1126/science.aaz9761

Helenius J, Brouhard G, Kalaidzidis Y, Diez S, Howard J (2006) The depolymerising kinesin MCAK uses lattice diffusion to rapidly target microtubule ends. Nature 441:115–119

Hiraizumi M, Yamashita K, Nishizawa T, Nureki O (2019) Cryo-EM structures capture the transport cycle of the P4-ATPase flippase. Science 365(6458):1149–1155. https://doi.org/10.1126/science.aay3353

van den Hoek H, Klena N, Jordan MA, Alvarez Viar G, Righetto RD, Schaffer M, Erdmann PS, Wan W, Geimer S, Plitzko JM, Baumeister W, Pigino G, Hamel V, Guichard P, Engel BD (2022) In situ architecture of the ciliary base reveals the stepwise assembly of intraflagellar transport trains. Science. 377(6605):543–548. https://doi.org/10.1126/science.abm6704

Huang L, Agrawal T, Zhu G, Yu S, Tao L, Lin JB, Marmorstein R, Shorter J, Yang X (2021) DAXX represents a new type of protein-folding enabler. Nature 597(7874):132–137. https://doi.org/10.1038/s41586-021-03824-5

Hussey PJ (2002) Microtubules do the twist. Nature 417:128–129

Huxley HE (1998) Getting to grips with contraction: the interplay of structure and biochemistry TIPS 23:84–87

Ibusuki R, Morishita T, Furuta A, Nakayama S, Yoshio M, Kojima H, Oiwa K, Furuta K (2022) Programmable molecular transport achieved by engineering protein motors to move on DNA nanotubes. Science. 375(6585):1159–1164. https://doi.org/10.1126/science.abj5170

Jenni S, Leibundgut M, Maier T, Ban N (2006) Architecture of a fungal fatty acid synthase at 5 Å resolution. Science 311:1263–1267

Junge W, Müller DJ (2011) Seeing a molecular motor at work. Science 333:704–705

Kaan HYK, Hackney DD, Kozielski F (2011) The structure of the kinesin-1 motor tail complex reveals the mechanism of autoinhibition. Science 333:883–885

Kalia R, Wang RY-R, Yusuf A, Thomas PV, Agard DA, Shaw JM, Frost A (2018) Structural basis of mitochondrial receptor binding and constriction by DRP1. Nature 558(7710):401–405. https://doi.org/10.1038/s41586-018-0211-2

Kang Y, Liu R, Wu J-X, Chen L (2019) Structural insights into the mechanism of human soluble guanylate cyclase. Nature 574(7777):206–210. https://doi.org/10.1038/s41586-019-1584-6

Kerssemakers JWJ, Munteanu EL, Laan L, Noetzel TL, Janson ME, Dogterom M (2006) Assembly dynamics of microtubules at molecular resolution. Nature 442:709–712

Kim YS, Kato HE, Yamashita K, Ito S, Inoue K, Ramakrishnan C, Fenno LE, Evans KE, Paggi JM, Dror RO, Kandori H, Kobilka BK, Deisseroth K (2018) Crystal structure of the natural anion-conducting channelrhodopsin GtACR1. Nature. 561(7723):343–348. https://doi.org/10.1038/s41586-018-0511-6

Kschonsak M, Chua HC, Weidling C, Chakouri N, Noland CL, Schott K, Chang T, Tam C, Patel N, Arthur CP, Leitner A, Ben-Johny M, Ciferri C, Pless SA, Payandeh J (2022) Structural architecture of the human NALCN channelosome. Nature. 603(7899):180–186. https://doi.org/10.1038/s41586-021-04313-5

Kueh HY, Mitchison TJ (2009) Structural plasticity in actin and tubulin polymer dynamics. Science 325:960–963

Kumar P, Paterson NG, Clayden J, Woolfson DN (2022) De novo design of discrete, stable $3_{10}$-helix peptide assemblies. Nature 607(7918):387–392. https://doi.org/10.1038/s41586-022-04868-x

Langan RA, Boyken SE, Ng AH, Samson JA, Dods G, Westbrook AM, Nguyen TH, Lajoie MJ, Chen Z, Berger S, Mulligan VK, Dueber JE, Novak WRP, El-Samad H, Baker D (2019) De novo design of bioactive protein switches. Nature. 572(7768):205–210. https://doi.org/10.1038/s41586-019-1432-8

Law RHP, Lukoyanova N, Voskoboinik I, Caradoc-Davies TT, Baran K, Dunstone MA, D'Angelo ME, Orlova EV, Coulibaly F, Verschoor S, Browne KA, Ciccone A, Kuiper MJ, Bird PI, Trapani JA, Saibil HR, Whisstock JC (2010) The structural basis for membrane binding and pore formation by lymphocyte perforin. Nature 468:447–451

Lee A, Hudson AR, Shiwarski DJ, Tashman JW, Hinton TJ, Yerneni S, Bliley JM, Campbell PG, Feinberg AW (2019) 3D bioprinting of collagen to rebuild components of the human heart. Science 365(6452):482–487. https://doi.org/10.1126/science.aav9051

Levin PA (2022) A bacterium that is not a microbe. Science 376(6600):1379–1380. https://doi.org/10.1126/science.adc9387

Liu Z, Zhang Z (2022) Mapping cell types across human tissues - Single-cell analyses reveal tissue-agnostic features and tissue-specific cell states. Science 376(6594):695–696. https://doi.org/10.1126/science.abq21

Lodish H, Berk A, Zipursky SL, Matsudaira P, Baltimore D, Darnell J (1999) Molecular cell biology. W. H. Freeman

Lord C, Bhandari D, Menon S, Ghassemian M, Nycz D, Hay J, Ghosh P, Ferro-Novick S (2011) Sequential interactions with Sec23 control the direction of vesicle traffic. Nature 473:181–186

Lovelock SL, Crawshaw R, Basler S, Levy C, Baker D, Hilvert D, Green AP (2022) The road to fully programmable protein catalysis. Nature 606(7912):49–58. https://doi.org/10.1038/s41586-022-04456-z

Lu P, Min D, DiMaio F, Wei KY, Vahey MD, Boyken SE, Chen Z, Fallas JA, Ueda G, Sheffler W, Mulligan VK, Xu W, Bowie JU, Baker D (2018) Accurate computational design of multipass transmembrane proteins. Science. 359(6379):1042–1046. https://doi.org/10.1126/science.aaq1739

Maier T, Jenni S, Ban N (2006) Architecture of mammalian fatty acid synthase at 4.5 Å resolution. Science 311:1258–1262

Maier T, Leibundgut M, Ban N (2008) The crystal structure of a mammalian fatty acid synthase. Science 321:1315–1322

Meldolesi J, Pozzan T (1998) The endoplasmic reticulum $Ca^+$ store: a view from the lumen. TIBS 23:10–14

Mitter M, Gasser C, Takacs Z, Langer CCH, Tang W, Jessberger G, Beales CT, Neuner E, Ameres SL, Peters JM, Goloborodko A, Micura R, Gerlich DW (2020) Conformation of sister chromatids in the replicated human genome. Nature. 586(7827):139–144. https://doi.org/10.1038/s41586-020-2744-4

Nakaseko Y, Yanagida M (2001) Cytoskeleton in the cell cycle. Nature 412:291–292

Nature (2021) 593(7860):570–574. https://doi.org/10.1038/s41586-021-03522-2

Navizet I, Lavery R, Jernigan RL (2004) Myosin flexibility: structural domains and collective vibrations. Proetisn Struct Func Bioinfo 54:384–393

Perez C, Gerber S, Boilevin J, Bucher M, Darbre T, Aebi M, Reymond JL, Locher KP (2015) Structure and mechanism of an active lipid-linked oligosaccharide flippase. Nature. 524(7566):433–438. https://doi.org/10.1038/nature14953

Pollard TD, Cooper JA (2009) Actin, a central player in cell shape and movement. Science 326:1208–1212

Prywer J (2022) The fascinating world of biogenic crystals. Science 376(6590):240–241. https://doi.org/10.1126/science.abo2781

Reddy AS, Warshaviak DT, Chachisvilis M (2012) Effect of membrane tension on the physical properties of DOPC lipid bilayer membrane. Biochim Biophys Acta 1818(9):2271–2281. https://doi.org/10.1016/j.bbamem.2012.05.00

Salway JG (2012) Medical biochemistry at a glance. Wiley-Blackwell

Schmidt AA (2002) The making of a vesicle. Nature 419:347–349

Service RF (2022) Tiny labmade motors are poised to do useful work. Science 376(6590):233. https://doi.org/10.1126/science.abq4278

Smith S (2006) Architectural options for a fatty acid synthase. Science 311:1251–1252

So WY, Tanner K (2022) The material properties of chromatin in vivo. Science 377(6605):472–473. https://doi.org/10.1126/science.add5444

Sui X, Wang K, Gluchowski NL, Elliott SD, Liao M, Walther TC, Farese RV Jr (2020) Structure and catalytic mechanism of a human triacylglycerol-synthesis enzyme. Nature 581(7808):323–328. https://doi.org/10.1038/s41586-020-2289-6

Tan L, Xing D, Chang CH, Li H, Xie XS (2018) Three-dimensional genome structures of single diploid human cells. Science 361(6405):924–928. https://doi.org/10.1126/science.aat5641

Tatham AS, Shewry PR (2000) Elastomeric proteins: biological roles, structures and mechanisms. TIBS 25:567–571

Viel A, Branton D (1996) Spectrin: on the path from structure to function. Curr Opin Cell Biol 8:49–55

Weis K (1998) Importins and exportins: how to get in and out of the nucleus. TIBS 23:185–189

Williams PM, Fowler SB, Best RB, Toca-Herrera JL, Scott KA, Steward A, Clarke J (2003) Hidden complexity in the mechanical properties of titin. Nature 422:446–449

Wintterlin J (2022) Growing polymers, caught in the act. Science 375(6585):1092–1093. https://doi.org/10.1126/science.abo2194

Wu X, Delbianco M, Anggara K, Michnowicz T, Pardo-Vargas A, Bharate P, Sen S, Pristl M, Rauschenbach S, Schlickum U, Abb S, Seeberger PH, Kern K (2020) Imaging single glycans. Nature. 582(7812):375–378. https://doi.org/10.1038/s41586-020-2362-1

Xu C, Lu P, Gamal El-Din TM, Pei XY, Johnson MC, Uyeda A, Bick MJ, Xu Q, Jiang D, Bai H, Reggiano G, Hsia Y, Brunette TJ, Dou J, Ma D, Lynch EM, Boyken SE, Huang P-S, Stewart L, DiMaio F, Kollman JM, Luisi BF, Matsuura T, Catterall WA, Baker D (2020) Computational design of transmembrane pores. Nature 585(7823):129–134. https://doi.org/10.1038/s41586-020-2646-5

Yamauchi Y, Helenius A (2013) Virus entry at a glance. J Cell Sci 126(Pt 6):1289–1295. https://doi.org/10.1242/jcs.119685

# Energy in Biology: Demand and Use

**2**

*A coupled energy source is a prerequisite of sustained dynamics in thermodynamically open systems.*

## Abstract

From the point of view of energy management in biological systems, a fundamental requirement is to ensure spontaneity. **Process spontaneity** is necessary since in a thermodynamically open system—such as the living cell—only spontaneous reactions can be catalyzed by enzymes. Note that enzymes do not, by themselves, contribute additional energy. Spontaneity of biological processes may be expressed by the following correlation, $\Delta G = \Delta H - T\Delta S$, where $\Delta G$ means the change of free energy; $\Delta H$, change of enthalpy; $\Delta S$, change of entropy; and $T$, temperature. Desirable processes which do not occur on their own must be coupled to other highly spontaneous mechanisms serving as energy sources. In biology, the fundamental sources of energy involve synthesis of water and photosynthesis. Since both processes are rather complex and cannot be exploited directly, they are used to synthesize ATP which acts as an energy carrier. Approaching biology from the point of view of elementary physics and chemistry reveals important mechanisms and enhances our understanding of various phenomena.

## Keywords

Spontaneity · Source of energy · Entropy-driven processes · Enthalpy-driven processes · Direct and indirect use of energy

1. Process spontaneity is denoted by $\Delta G$. What does $\Delta G_0$ stand for?
2. How can non-spontaneous processes be imbued with spontaneity?
3. How does the $\Delta G$ parameter relate to cell life and cell death?
4. How does water synthesis drive the synthesis of phosphoanhydride bonds in ATP?
5. What is the role of intermembrane space in mitochondria?
6. What are the direct and indirect products of photosynthesis?
7. Energy carriers—their role and importance.
8. Why aren't biological processes 100% efficient?
9. Processes dominated by entropy—examples.
10. How do organisms solve the problem of limited availability of oxygen during muscle contractions?
11. The kinetics of muscle activity.

## 2.1    General Principles of Thermodynamics

Physical and chemical processes may only occur spontaneously if they generate energy or non-spontaneously if they consume it. However, all processes occurring in a cell must have a spontaneous character because only these processes may be catalyzed by enzymes. Enzymes merely accelerate reactions; they do not provide energy.

In the inanimate world, non-spontaneous (endergonic) reactions, including most synthesis processes, consume thermal energy. In a cell, chemical energy can be derived from exergonic (energy-producing) processes. An important source of energy in living organisms is sunlight—the driving force in photosynthesis.

Due to high susceptibility of living organisms to heat damage, thermal energy is inconvenient.

Catalysis of inherently non-spontaneous processes becomes possible only when they are thermodynamically coupled to other spontaneous processes in such a way that the resulting complex process dissipates energy.

Examples of inherently non-spontaneous processes which acquire spontaneity by relying on exergonic reactions include:

1. Synthesis
2. Structural rearrangement of proteins (e.g., in muscle contraction)

Processes related to degradation are usually spontaneous by nature, and most of their stages do not require additional exergonic processes as a source of energy.

In physical terms, spontaneity is subject to Gibb's definition, where $\Delta G$ (change in free energy) corresponds to $\Delta H$ (change in enthalpy) and $\Delta S$ (change in entropy), according to the following equation:

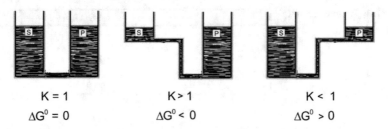

$$K = 1 \qquad\qquad K > 1 \qquad\qquad K < 1$$
$$\Delta G^0 = 0 \qquad\qquad \Delta G^0 < 0 \qquad\qquad \Delta G^0 > 0$$

**Fig. 2.1**   The relationship between $\Delta G^0$ and the equilibrium constant $K$ in a model system

$$\Delta G = \Delta H - T\Delta S$$

($T$—temperature in $°K$)

The change in enthalpy associated with a chemical process may be calculated as a net difference in the sum of molecular binding energies prior to and following the reaction.

Entropy is a measure of the likelihood that a physical system will enter a given state. Since chaotic distribution of elements is considered the most probable, physical systems exhibit a general tendency to gravitate toward chaos. Any form of ordering is thermodynamically disadvantageous.

According to the presented formula, energy loss and the corresponding increase in entropy ($\Delta H$ and $\Delta S$) are the hallmarks of a spontaneous process.

For specific processes, the change in free energy may be determined without referring to the presented mechanism and instead relying on reaction dynamics; specifically, on the ratio of product and substrate concentrations:

$$\Delta G = \Delta G^0 + RT^* \, ln\{([C] * [D])/([A] * [B])\}$$

$R$, gas constant; $T$, absolute temperature.

The $\Delta G^0$ parameter is a measure of spontaneity. It depends on the properties of process elements and is therefore a function of the state of equilibrium. It may be derived from the ratio of product and substrate concentrations once a state of equilibrium has been reached.

If the process is in equilibrium, $\Delta G$ becomes equal to 0, and thus, according to the formula, $\Delta G^0$ is given as

$$\Delta G^0 = -RT\{ ln \, (K)\}$$

where $K$ is the equilibrium constant (Fig. 2.1 and Table 2.1).

The figure depicts three types of communicating vessels in which the liquid has reached a state of equilibrium corresponding to various ratios of "products" (right vessel) and "substrates" (left vessel). The table (in Fig. 2.1) presents some numerical examples of the relation between $K$ and $\Delta G^0$.

The true measure of spontaneity is therefore not $\Delta G^0$ but $\Delta G$, which expresses the capability to perform work for a reaction which is not in a state of equilibrium. In

**Table 2.1**  An integral part of Fig. 2.1

| $K$ | $\Delta G^0$ (kJ/mol)—temp. 25 °C |
|---|---|
| $10^4$ | −22.8 |
| $10^1$ | −5.7 |
| $10^0$ | 0.0 |
| $10^{-1}$ | 5.7 |
| $10^{-4}$ | 22.8 |

contrast, $\Delta G^0$ merely indicates the reactivity of substrates as a consequence of their physical and chemical nature.

A reaction which is in a state of equilibrium cannot perform useful work. Energy can only be extracted from processes which have not yet reached equilibrium.

Maintaining a steady state of nonequilibrium is possible only in a thermodynamically open environment where energy and reaction components may flow through the system. This, in turn, calls for an external source of energy as well as a means of automatic control (see Fig. 1, Introduction).

Any spontaneous reaction can be treated as a source of energy as long as its spontaneity is sufficient for the thermodynamically disadvantageous reaction to occur and provided that both processes are thermodynamically coupled. In practice, synthesis reactions may only draw energy from highly spontaneous processes due to the need to form covalent bonds.

The chemical reactions which power biological processes are characterized by varying degrees of efficiency. In general, they tend to be on the lower end of the efficiency spectrum, compared to energy sources which drive matter transformation processes in our universe.

In search for a common criterion to describe the efficiency of various energy sources, we can refer to the net loss of mass associated with a release of energy, according to Einstein's formula:

$$E = m\,c^2$$

The $\Delta M/M$ coefficient (relative loss of mass, given, e.g., in %) allows us to compare the efficiency of energy sources. The most efficient processes are those involved in the gravitational collapse of stars. Their efficiency may reach 40%, which means that 40% of the stationary mass of the system is converted into energy. In comparison, nuclear reactions have an approximate efficiency of 0.8%.

The efficiency of chemical energy sources available to biological systems is incomparably lower and amounts to approximately $10^{-7}$% (Fig. 2.2).

Among chemical reactions, the most potent sources of energy are found in oxidation processes, commonly exploited by biological systems. Oxidation tends to result in the largest net release of energy per unit of mass, although the efficiency of specific types of oxidation varies. For instance, the efficiency of hydrogen-halogen reactions (expressed in kcal/mol), calculated as the balance of binding energies, is as follows:

**Fig. 2.2** Pictorial
representation of the
efficiency of selected energy
sources, including stellar
collapse, nuclear reactions,
and chemical processes

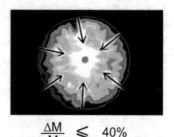

$$\frac{\Delta M}{M} \leqslant 40\%$$

$$\frac{\Delta M}{M} = 0{,}8\%$$

$$\frac{\Delta M}{M} = 10^{-7}\%$$

$$H_2 + F_2 = 2HF \quad 104 + 37 = 2^*135(\text{kcal/mol}) \quad \Delta H = 129(\text{kcal/mol})$$

$$H_2 + Cl_2 = 2HCl \quad 104 + 58 = 2^*103(\text{kcal/mol}) \quad \Delta H = 44(\text{kcal/mol})$$

$$H_2 + Br_2 = 2HBr \quad 104 + 46 = 2^*87.5(\text{kcal/mol}) \quad \Delta H = 25(\text{kcal/mol})$$

Under similar conditions the reaction between hydrogen and oxygen yields an average of 56.7 kcal/mol. It should come as no surprise that—given unrestricted access to atmospheric oxygen and to hydrogen atoms derived from hydrocarbons—the combustion of hydrogen (i.e., the synthesis of water; $H_2 + 1/2O_2 = H_2O$) has become a principal source of energy in nature, next to photosynthesis, which exploits the energy of solar radiation.

## 2.2    Biological Energy Sources: Synthesis of Water

Hydrogen is combined with oxygen on inner mitochondrial membrane, while the conversion of hydrogen carriers (lipids, sugars, and proteins) into forms appropriate for water synthesis occurs in the cytoplasm and the mitochondrial matrix. Major energy-generating metabolic pathways include glycolysis, β-oxidation, and degradation of amino acids.

The basic process associated with the release of hydrogen and its subsequent oxidation (called the Krebs cycle) is augmented by processes which transfer electrons onto oxygen atoms (Fig. 2.3).

Oxidation occurs in stages, enabling optimal use of the released energy. An important byproduct of water synthesis is the universal energy carrier known as ATP (synthesized separately).

As water synthesis is a highly spontaneous process, it can be exploited to cover the energy debt incurred by endergonic synthesis of ATP, as long as both processes are thermodynamically coupled, enabling spontaneous catalysis of anhydride bonds in ATP.

Water synthesis is a universal source of energy in heterotrophic systems. In contrast, autotrophic organisms rely on the energy of light which is exploited in

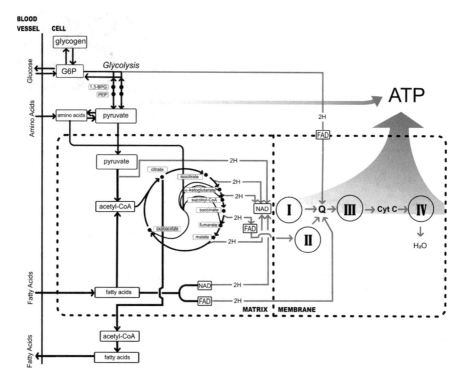

**Fig. 2.3**  A schematic depiction of energy conversion processes in living cells

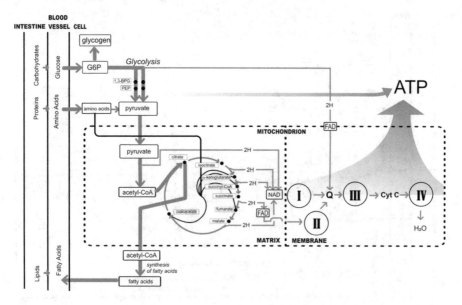

**Fig. 2.4** Directed energy transfer occurring in a living cell when nutrients (hydrogen carriers) are freely available (fed state)

the process of photosynthesis. Both processes yield ATP along with reduced pyridine nucleotides.

As mentioned above, linking the spontaneous process of water synthesis with non-spontaneous creation of anhydride bonds in ATP is a prerequisite for achieving a thermodynamically unified system. This is done by introducing a hydrogen ion gradient which affects both processes, although in different ways:

A. The respiratory chain—by carrier proteins which transfer hydrogen atoms and electrons and must therefore be able to dissociate or attach hydrogen ions, enabling their transduction across the mitochondrial membrane and giving rise to an ion gradient
B. ATP synthesis—by exploiting the energy released in the spontaneous discharge of the hydrogen ion gradient

The hydrogen atoms used to synthesize water in the respiratory chain are derived from nutrients and can directly participate in the chain by way of the Krebs cycle (TCA). Nutrients include sugars (mostly glucose), amino acids, and lipids, which reach hepatocytes following absorption from the small intestine. The energy they carry is exploited in sequestration processes (mainly fatty acid synthesis and lipogenesis) (Fig. 2.4).

During periods of starvation, hydrogen carriers can be retrieved from storage: glucose comes from glycogen, while fatty acids are extracted from adipose tissue.

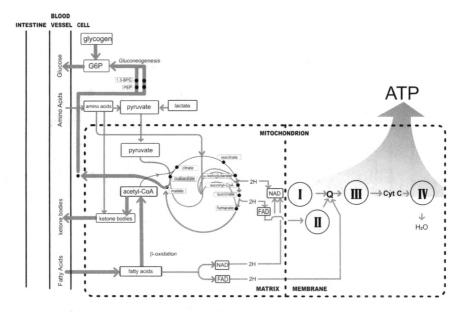

**Fig. 2.5** Directed energy transfer occurring in a living cell whenever stored resources must be expended to maintain appropriate levels of glucose in blood (fasted state)

Under conditions of high physical exertion or inadequate food intake, the extraction of lipids increases, and most of the released energy is used to resynthesize glucose from amino acids (Fig. 2.5).

Dehydrogenation of substrates is catalyzed by pyridine- and flavin-linked dehydrogenases. Respiratory chain proteins are integrated in the mitochondrial membrane transport hydrogen atoms and electrons onto oxygen ($O_2$) molecules across the potential gradient, via complexes I, III, and IV, namely, NADH dehydrogenase (complex I) and coenzyme Q, cytochrome $bc_1$ complex reductase (complex III), cytochrome c, and finally cytochrome c oxidase (complex IV). Hydrogen atoms released by dehydrogenation of succinate are introduced to the respiratory chain via complex II (succinate dehydrogenase) and coenzyme Q, skipping complex I (Figs. 2.3, 2.4, and 2.5).

Complexes I, III, and IV are integrated in the membrane. In contrast, coenzyme Q (mediating capture of electrons by complex III) and cytochrome c are mobile, although the former is located within the membrane, while the latter is found in the intermembrane space. The double-electron NAD dehydrogenase initiator changes to a conduit for univalent electron carriers (iron-sulfur proteins, cytochromes) and a four-electron channel in the last phase of oxygen reduction.

Water synthesis ultimately results in phosphorylation, which (in this case) converts ADP to ATP. Since an anhydride bond must be created, there is a need of energy. ATP synthesis must therefore be coupled to the synthesis of water if spontaneity is to be maintained. The link is effected by introducing a hydrogen ion gradient between the mitochondrial matrix and the intermembrane space, which

**Fig. 2.6** Various forms of protein-membrane integration (gray zones indicate hydrophobic surfaces)

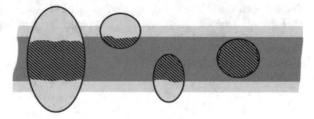

requires ejecting of hydrogen ions into the intermembrane space in the course of electron transportation for synthesis of water. The structures responsible for transporting ions must do so in a predetermined direction and maintain functional specificity. This task, like many others, is performed by dedicated proteins.

Ion transport is usually effected by a protein-specific change in pK of selected proton-binding groups, similar to the Bohr effect which occurs in hemoglobin. Transduction of hydrogen ions from the mitochondrial matrix to the intermembrane space works against the emerging ion gradient. Thus, protein ion channels must also fulfill the role of a sluice gate. Electron carriers participating in the highly spontaneous process of water synthesis in the membrane also act as transverse carriers of hydrogen ions. Thus owing to the mutual dependence of both ways of transport, the non-spontaneous formation of an ion gradient may draw energy from the oxidation process.

A suitable direction of transduction is ensured by maintaining proper alignment and integration of hydrogen and electron carrier proteins in the membrane so that protons are captured and released on specific sides (either within the mitochondrial matrix or in the intermembrane space). The structure of membrane proteins and their localization in the membrane is well suited to this task. Their apolar amino acids and their shape enforce the correct alignment and integration of protein molecules in the membrane (Fig. 2.6).

The ability to transport hydrogen ions is a property of proteins forming complexes I and IV. In complex III a similar function is most likely performed by coenzyme Q. Its specific structure and integration with the membrane, as well as its interaction with proteins, ensure unidirectional ion transfer. Coenzyme Q is an apolar, non-protein mobile molecule, consisting of a quinone derivative ring and a polyisoprenyl chain which, in humans, contains ten elements. The carbonyl groups of the quinone ring may undergo reduction by a hydrogen ion or by an electron (Fig. 2.7).

A proton may also dissociate from the hydroxyl group, leaving behind an anionic residue.

According to the most widely accepted hypotheses, the ubiquinone molecule (coenzyme Q) is able to rotate in the hydrophobic area of the membrane, coming into contact with integrated respiratory chain proteins and mediating the transduction of electrons and hydrogen ions. Through reduction (with an electron), the carboxyl group assumes a polar form and migrates from the membrane to the aqueous environment where it immediately attracts a proton (as its pK precludes the existence

**Fig. 2.7**  Ubiquinone
oxidation and reduction
products: (**a**) ubiquinone, (**b**)
semiquinone intermediate, (**c**)
ubiquinol, and (**d**) dissociated
ubiquinol

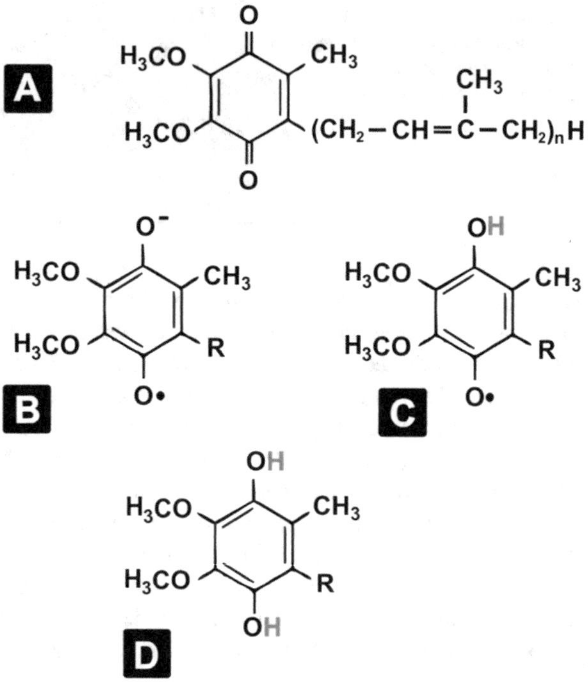

of a dissociated form in a non-alkaline environment). Subsequently, the ubiquinone
again becomes apolar and returns to the hydrophobic zone of the membrane. While
rotating it encounters cytochrome b and releases an electron, converting to
semiquinone or quinone.

The presence of iron-sulfur proteins within the mitochondrial matrix, and of
cytochrome b in the intermembrane space, results in a situation where the shortest
proton transport route across the membrane is the one provided by ubiquinone. The
source of energy powering this process is the electron transport carried by the
respiratory chain (Fig. 2.8). A suitable arrangement of respiratory chain proteins is
therefore crucial for coupling ATP synthesis to the synthesis of water.

The process depends on respiratory chain proteins being integrated and properly
aligned in the membrane (either on the side of the matrix or in the intermembrane
space) as well as on the presence of mobile ubiquinone molecules.

The arrangement of proteins which participate in binding hydrogen ions and
transporting electrons across the membrane is such that electron transduction follows
changes in the oxidation-reduction potential. A proof of structural ordering can be
found in the stability of complexes I, II, III, and IV, even in their isolated forms. All
enzymes connected with hydrogen and electron transportation are permanently
integrated in the membrane with exception of few TCA cycle dehydrogenases and
cytochrome c. For obvious reasons, this does not apply to NADH dehydrogenases
which participate in the Krebs cycle in the mitochondrial matrix. Also the loose bond
between the membrane and cytochrome c, facilitating its mobility and ability to bind

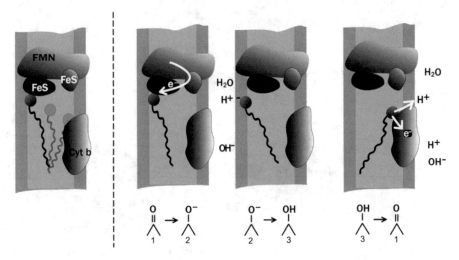

**Fig. 2.8** Hypothetical hydrogen ion transduction mechanism between the mitochondrial matrix and the intermembrane space with the participation of ubiquinone

electrons, is most likely maintained in order to mask variations in the availability of oxygen while simultaneously enabling the transfer of concerted transfer of four electrons onto an oxygen molecule with the aid of cytochrome c oxidase:

$$4 \text{ cyt c red} + 4\text{H}^+ + \text{O}_2 = > 4 \text{ cyt.c ox} + 2\text{H}_2\text{O}$$

Cytochrome c molecules, loosely integrated with the membrane on its dorsal surface (i.e., within the intermembrane space), may therefore act as an electron reservoir, maintaining the consistency of the respiratory chain.

## 2.3    ATP Synthesis

The structure responsible for creating a high-energy anhydride bond in the process of converting ADP to ATP is a dedicated protein complex called ATP synthase—a mushroom-shaped structure anchored in the inner mitochondrial membrane. It is a very efficient machine, drawing power from the hydrogen ion gradient which itself emerges as a result of water synthesis. In order to achieve spontaneous synthesis of an anhydride bond, discharge of the ion gradient must release significantly more energy than the ATP synthesis process consumes. The synthase complex undergoes rotation which results in reordering of its structural components. It is capable of synthesizing approximately 100 ATP molecules per second, discharging three protons per molecule, although its specific mechanism of action remains still unclear.

It seems valid to assume that changes in the structure of the synthase effected by the ion gradient should result in shifting the ADP substrate (bound to the active site)

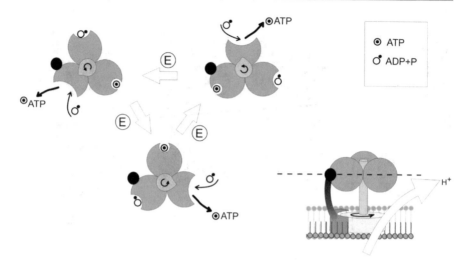

**Fig. 2.9** Simplified diagram of ATP synthesis

and the phosphate group to an environment characterized by low polarity. According to this theory, the ATP molecule, which is less polar than the substrates used in its synthesis (ADP and P) and which tends to undergo spontaneous hydrolysis in the presence of water, becomes now easier to synthesize following a change in environmental conditions. Thus, a change in polarity driven by synthesis of water may reverse the direction of spontaneity making the process of ATP synthesis spontaneous.

An alternative hypothesis assumes that the reversal in the spontaneity of ATP hydrolysis is a result of local acidification. It can be observed that the rate of ATP hydrolysis decreases rapidly in low-pH environments. The high susceptibility of ATP to pH changes in the neutral space is a consequence of changes in pK of the -OH moiety, which migrates from an alkaline area (in a free phosphate group) to an area with pH = 6.95 (in ATP).

It is also likely that both mechanisms (reversal in polarity and acidification) come into play simultaneously. Observational evidence seems to propose even more convincing theory suggesting that the structural rearrangement of synthase components results in the creation of a binding site with high affinity for ATP, far exceeding its corresponding affinity for ADP. This effect negates the spontaneity of ATP hydrolysis; however it also introduces a very strong bond between the enzyme and the ATP molecule, making it difficult to release the product. This is why the process requires a source of motive power (the aforementioned ion gradient discharge), driving structural changes in the synthase complex. Once the gradient is restored, the synthesis process may repeat itself (Fig. 2.9).

The central rotating element (subunit γ, marked in color) forces a realignment of non-rotating catalytic subunits. Rotation is powered by the hydrogen ion gradient discharge. Side view and cross-section are presented.

**Fig. 2.10** Coenzyme A—the grayed-out area indicates the acetyl residue

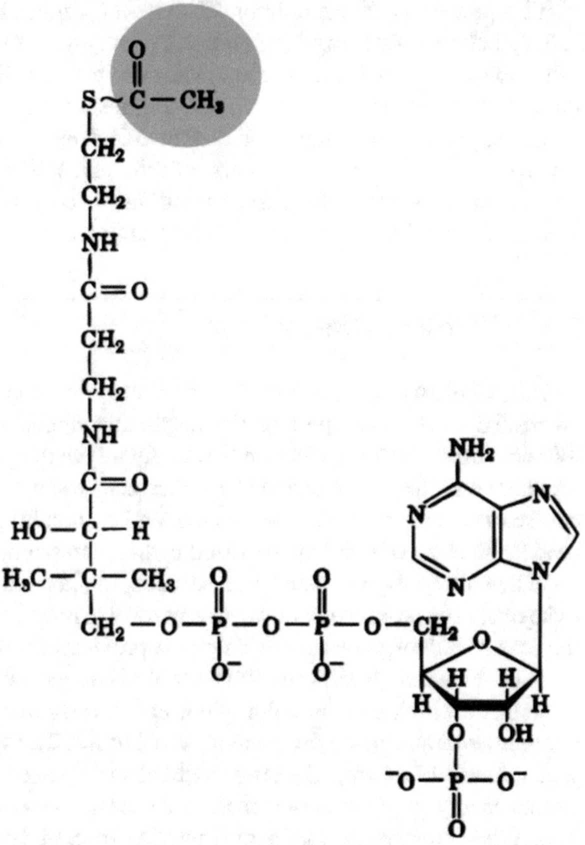

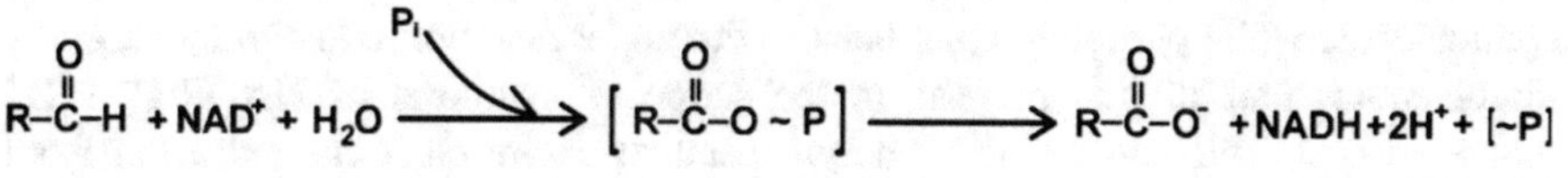

**Fig. 2.11** Creation of a high-energy bond (mixed anhydride) through carbon oxidation (glyceraldehyde 3-phosphate oxidation)

Preparing nutrients (hydrogen carriers) for participation in water synthesis follows different paths for sugars, lipids, and proteins. This is perhaps obvious given their relative structural differences; however, in all cases the final form, which acts as a substrate for dehydrogenases, is acetyl-CoA (Fig. 2.10).

Sugars are converted into acetyl-CoA in the glycolysis process, following oxidative pyruvate decarboxylation; fatty acids undergo β-oxidation, while proteins are subject to hydrolysis, deamination, and some changes in their hydrocarbon core.

Aside from water synthesis, part of the energy required to form high-energy bonds comes from oxygenation of carbon atoms. The glycolysis process affords two molecules of ATP as a result of glyceraldehyde 3-phosphate oxidation (Fig. 2.11).

This process is an example of substrate-level phosphorylation and occurs in the cell cytoplasm. As a result, high-energy anhydride (1,3-biphosphoglycerate) and ester bonds (phosphoenolpyruvate acid) are created directly within the substrate, through dehydrogenation and dehydration, respectively.

Oxidation of carbon atoms yields acetyl-CoA as a product of the thiolysis process involved in β-oxidative degradation of fatty acids. The energy gain resulting from carbon oxidation covers the cost of synthesis of acetyl-CoA and succinyl-CoA in the course of oxidative decarboxylation of α-keto acids.

## 2.4    Photosynthesis

Photosynthesis is a process which—from the point of view of electron transfer—can be treated as a counterpart of the respiratory chain. In heterotrophic organisms, mitochondria transport electrons from hydrogenated compounds (sugars, lipids, proteins) onto oxygen molecules, synthesizing water in the process, whereas in the course of photosynthesis, electrons released by breaking down water molecules are used as a means of reducing oxidized carbon compounds (Fig. 2.12).

In heterotrophic organisms the respiratory chain has a spontaneous quality (owing to its oxidative properties); however any reverse process requires energy to occur. In the case of photosynthesis, this energy is provided by sunlight (Fig. 2.13).

The reduction process consumes solar energy by way of dedicated protein "antennae" which contain chlorophyll and carotenoids, each capable of capturing photons and transferring their energy to electrons. The system resembles the macroscopic model of a pumped-storage hydroelectric power plant (Fig. 2.14) which uses excess energy to pump downstream water back to its upper reservoir during nighttime. When the demand for energy increases (during daylight hours), water can flow back to the lower reservoir, generating power. Clearly, this direction of flow is spontaneous, while pumping water back to the upper reservoir consumes energy. A direct counterpart of this process in the scope of synthesis of $NADPH^+ + H^+$ reductors is its reliance on light energy, used to excite electrons extracted from water molecules.

Photosystems P680 and P700 perform the role of "electron pumps," imparting electrons with additional energy.

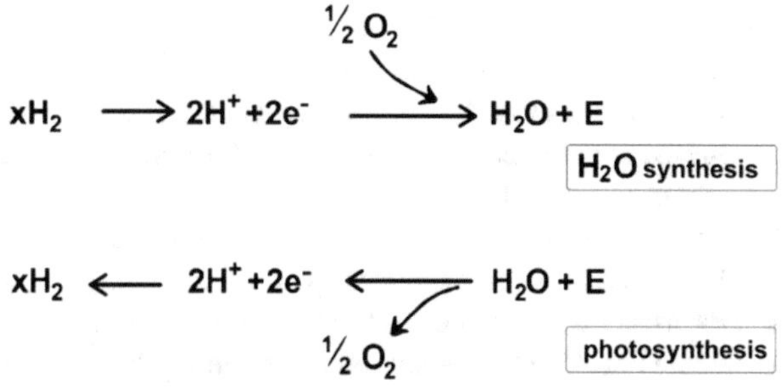

**Fig. 2.12** Synthesis of water using the hydrogen atoms released by hydrogenated compounds ($xH_2$) within the respiratory chain, compared to synthesis of hydrogenated compounds from water in the presence of solar energy (photosynthesis)

$$xH_2 \longrightarrow 2H^+ + 2e^- \xrightarrow{\frac{1}{2}O_2} H_2O + E$$

$$\boxed{H_2O \text{ synthesis}}$$

$$xH_2 \longleftarrow 2H^+ + 2e^- \longleftarrow H_2O + E$$

$$\frac{1}{2}O_2$$

$$\boxed{\text{photosynthesis}}$$

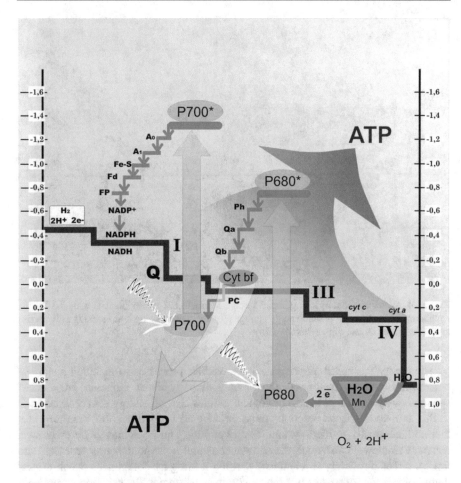

**Fig. 2.13** Stage-by-stage electron flow comparison between the respiratory chain (black descending path) and photosynthesis (gray pathway), with energy changes indicated. Ph, pheophytin; Qa and Qb, quinone carriers; cyt bf, cytochrome bf complex; Pc, plastocyanin; $A_0$ and $A_1$, electron acceptors; Fe-S, iron-sulfur center; Fd, ferredoxin; Fp, NADP reductase

Assimilation and reduction of carbon dioxide (dark stage) are facilitated by reduced NADPH nucleotides which are produced in the light-dependent stage.

Hydrogen combustion and photosynthesis are the basic sources of energy in the living world. As they are subject to common laws of physics, their operating principles resemble those observed in many macroscopic systems (Fig. 2.15).

## 2.5   Direct and Indirect Exploitation of Energy Sources

Direct exploitation (direct coupling of spontaneous and non-spontaneous processes to an energy source)

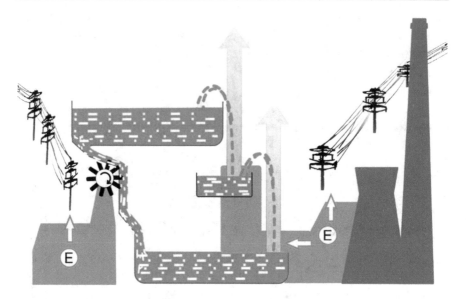

**Fig. 2.14** Water circulation in a pumped-storage hydroelectric power plant as a model for the circulation of electrons in natural energy storage systems (synthesizing and breaking down water molecules in the course of photosynthesis)

The ability to exploit an energy source enables processes to maintain a state of nonequilibrium. As mentioned above, only these types of processes may occur spontaneously and be of use to biological entities. For an energy source to become useful, non-spontaneous reactions must be coupled to its operation, resulting in a thermodynamically unified system. Such coupling can be achieved by creating a coherent framework in which the spontaneous and non-spontaneous processes are linked, either physically or chemically, using a bridging component which affects them both. If the properties of both reactions are different, the bridging component must also enable suitable adaptation and mediation. In a water mill, the millstone shaft couples water flow (which acts as an energy source) to the work performed by quern-stones. A car engine combusts gasoline, using the released energy to impart motive force to the wheels. In some situations both processes (the spontaneous one and the non-spontaneous one) may share similar characteristics—for instance, skiers may propel themselves down a slope in order to effortlessly ski up the opposite slope. In this case, both processes involve skiing and are therefore similar; however a prerequisite of coupling is that both slopes need to be in a close proximity to each other (otherwise the energy gained in the descent would be lost).

Direct exploitation of the energy released via the hydrolysis of ATP is possible usually by introducing an active binding carrier mediating the energy transfer.

Carriers are considered active as long as their concentration ensures a sufficient release of energy to synthesize a new chemical bond by way of a non-spontaneous process. Active carriers are relatively short-lived and exist either in the active site of an enzyme or as independent substrates. In the latter case, they can be treated as

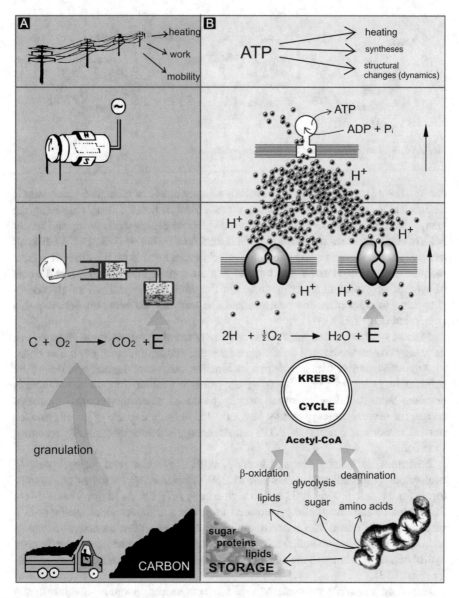

**Fig. 2.15** Comparison of corresponding stages in the synthesis of universal energy carriers: (**a**) physical processes (involving electricity) and (**b**) chemical processes (ATP synthesis), showing analogies between both approaches

distinct components of cellular metabolic pathways. Examples include PRPP, UDPG, active mediators of cholesterol synthesis, and others. If the phosphorylation resulting from ATP hydrolysis yields a low-energy compound (such as glucose-6-phosphate), the reaction may consist of a single stage, and its product is not an active

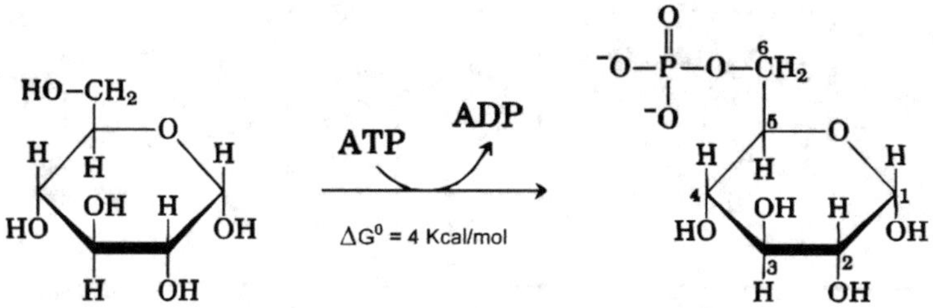

**Fig. 2.16** Spontaneous synthesis of glucose-6-phosphate

carrier. The energy released via hydrolysis of ester bonds in glucose-6-phosphate is insufficient to cover the cost of creating a new chemical bond. The net change in free energy ($\Delta G^{0}$) associated with hydrolysis of glucose-6-phosphate is approximately 3.3 kcal/mol, whereas synthesis costs 4.0 kcal/mol [$-7.3 - (-3.3) = -4.0$ kcal/mol]. Since the binding energy for a typical chemical bond is on the order of 3 kcal/mol, the energy stored in the ester bond of glucose-6-phosphate cannot cover the cost of synthesizing additional compounds. On the other hand, synthesis of glucose-6-phosphate is a spontaneous process, and the cell has no problem deriving this compound (Fig. 2.16).

In most synthesis reactions, ATP is only involved at an intermediate stage, where its energy covers the cost of creating a new ester, amide, thioester, or similar bond.

Any active carrier which performs its function outside of the active site must be sufficiently stable to avoid breaking up prior to participating in the synthesis reaction. Such mobile carriers are usually produced when the required synthesis consists of several stages or cannot be conducted in the active site of the enzyme for sterical reasons. Contrary to ATP, active energy carriers are usually reaction-specific.

Examples of active carriers include acid anhydrides and other types of compounds (especially thioesters and esters). Mobile energy carriers are usually formed as a result of hydrolysis of two high-energy ATP bonds. In many cases this is the minimum amount of energy required to power a reaction which synthesizes a single chemical bond. The adenosine residue often generates an active carrier in addition to a reaction product, while the dissociated pyrophosphate group undergoes hydrolysis, contributing to the spontaneity of the process and limiting its reversibility (Fig. 2.17).

Expelling a mobile or unstable reaction component in order to increase the spontaneity of active energy carrier synthesis is a process which occurs in many biological mechanisms, including decarboxylation of malonyl-CoA at the initial stage of fatty acid synthesis and decarboxylation of oxaloacetate acid at the initial stage gluconeogenesis, where an easily diffusing carbon dioxide molecule is ejected from the reaction site (Fig. 2.18).

The action of active energy carriers may be compared to a snowball rolling down a hill. The descending snowball gains sufficient energy to traverse another, smaller

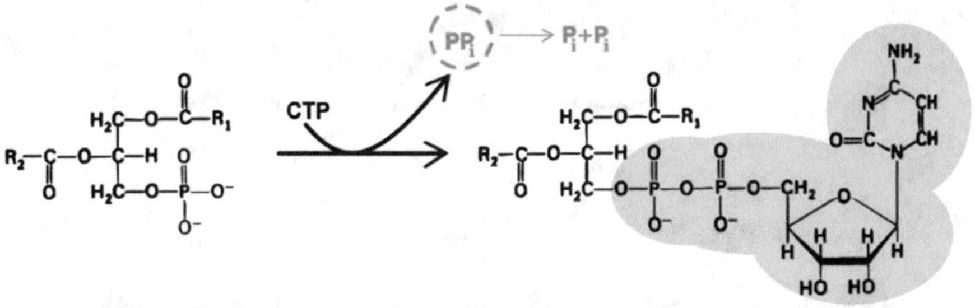

**Fig. 2.17** Promoting the irreversibility of active energy carrier synthesis via releasing and hydrolysis of pyrophosphates. Creation of an activated diacylglycerol molecule

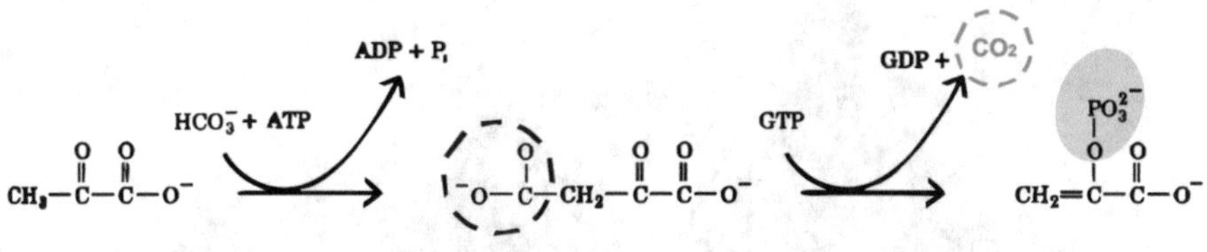

**Fig. 2.18** Ensuring irreversibility of a chemical process by dissociating easily diffunding carbon dioxide molecules: synthesis of phosphoenolpyruvate

mound, adjacent to its starting point. In our case, the smaller hill represents the final synthesis reaction (Fig. 2.19).

Common energy carriers include pyrophosphate containing compounds (cholesterol synthesis intermediates, UDP-glucose PRPP), carboxybiotin (active carbon dioxide), S-adenosyl methionine (active methyl group), and many others (Fig. 2.20).

If an active energy carrier is created and subsequently consumed to form a new bond within the active site of the enzyme, synthesis of that enzyme's product may become energetically advantageous. What is more, correct alignment of substrates in the active pocket increases their likelihood of coming into contact with each other and therefore contributes to the spontaneity of the synthesis reaction. To illustrate this process, let us consider the formation of an amide bond in glutamine. This reaction involves an active carrier (mixed acyl-phosphate anhydride), synthesized at the cost of one high-energy ATP bond (Fig. 2.21). In this case, the energy of the carrier is comparable to that of the source. A similar situation occurs in the synthesis of phosphocreatine.

Proper alignment of substrates in the active pocket ensures direct contact, mimicking increased concentrations of both substances.

$$ATP + \text{creatine} <=> ADP + \text{phosphocreatine}$$

The process is not inherently spontaneous ($\Delta G^0 = 3$ kcal/mol); however its direction can be determined by changes in the concentration of ATP in myocytes between work and rest periods. Thus, phosphocreatine acts as a reservoir of

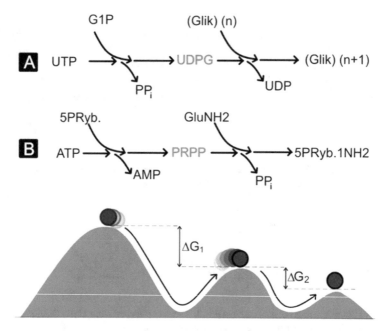

**Fig. 2.19** Schematic depiction of the synthesis of active energy carriers and their role as coupling factors, linking energy sources to product synthesis processes: (**a**) synthesis of glycogen (Glyc)$_{n+1}$ from (Glyc)$_n$ and (**b**) synthesis of 5-phosphorybosylamine

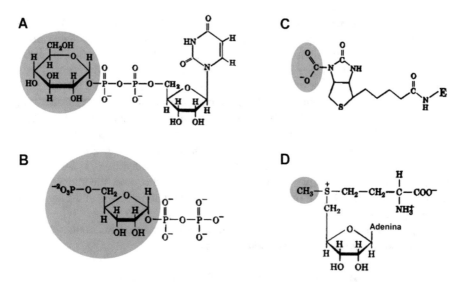

**Fig. 2.20** Examples of active energy carriers: (**a**) active glucose (UDPG), (**b**) active ribose (PRPP), (**c**) active carbon dioxide (carboxybiotin), and (**d**) active methyl group (S-adenosyl methionine)

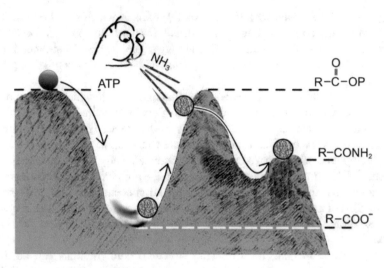

**Fig. 2.21** Synthesis of a glutamine amide bond in the active site as an example of using a carrier whose energy is comparable to that of the source

high-energy bonds, accumulating while the muscle is at rest and resynthesizing ATP when physical effort is required.

Understanding the role of active carriers is essential for the study of metabolic processes.

The second category of processes, directly dependent on energy sources, involves structural reconfiguration of proteins, which can be further differentiated into low- and high-energy reconfiguration. Low-energy reconfiguration occurs in proteins which form weak, easily reversible bonds with ligands. In such cases, structural changes are powered by the energy released in the creation of the complex. The scope of reconfiguration depends on the packing of the protein and on its stability. Packing a folded, single-chain native protein results in a relatively stable structure, immune to significant changes which may result from the creation of a noncovalent protein-ligand bond. As a rule, the protein assumes the structural configuration corresponding to its global energy minimum. However, it may also temporarily stabilize in a local minimum (if the associated energy is not significantly greater than that of the global minimum). In most cases, such an intermediate structure is inherently unstable, and the protein spontaneously refolds to its most stable configuration.

Important low-energy reconfiguration processes may occur in proteins which consist of subunits. Structural changes resulting from relative motion of subunits typically do not involve significant expenditures of energy. Of particular note are the so-called allosteric proteins—for instance, hemoglobin (hemoglobin T (tens) ↔ hemoglobin R (relaxed)), whose rearrangement is driven by a weak and reversible bond between the protein and an oxygen molecule. Allosteric proteins are genetically conditioned to possess two stable structural configurations, easily swapped as a result of binding or releasing ligands. Thus, they tend to have two comparable energy minima (separated by a low threshold), each of which may be treated as a global minimum corresponding to the native form of the protein. Given such properties, even a weakly interacting ligand may trigger significant structural reconfiguration. This phenomenon is of critical importance to a variety of regulatory proteins.

In many cases, however, the second potential minimum in which the protein may achieve relative stability is separated from the global minimum by a high threshold requiring a significant expenditure of energy to overcome. For proteins associated with motor functions or other difficult structural changes, the ligand must be a high-energy compound which "pays" for the thermodynamically disadvantageous structural change by forming an active complex with the protein. Contrary to low-energy reconfigurations, the relative difference in ligand concentrations is insufficient to cover the cost of a difficult structural change. Such processes are therefore coupled to highly exergonic reactions such as ATP hydrolysis. Figure 2.22 depicts this type of situation.

In the case of myosin, the following structural configurations can be distinguished: relaxed structure with no ligand present (protein molecule without ATP) (Fig. 2.22c1), ATP-bound structure (Fig. 2.22c2), and contracted structure resulting from hydrolysis of a high-energy bond (Fig. 2.22c3) prior to releasing the products of this reaction. Relaxation of myosin (Fig. 2.22c3 and c1) is a highly spontaneous process, capable of performing useful work (Figs. 2.22c4, 2.23, and 2.24). Phases 1–4 are depicted in Figs. 2.23, 2.24, and 2.25. Catalysis of structural changes by high-energy bonds is not limited to motor proteins—it can also be observed in ribosomes, microtubule synthesis, protein G functions, and many other biological processes.

**Indirect Method (Indirect Coupling Between Non-spontaneous Processes and Sources of Energy)**
The link between a biological process and an energy source does not have to be immediate. Indirect coupling occurs when the process is driven by relative changes in the concentration of reaction components. The cell—a thermodynamically open system—may take advantage of external energy sources by acquiring substrates and expelling reaction products. This model is similar to an apartment block where tenants may utilize power, water, and natural gas supplied ready to use to their building.

The equation

$$\Delta G = \Delta G^0 + RT^* Ln\{[C]^*[D])/([A]^*[B])\}$$

indicates that reactions may acquire spontaneity if substrate concentrations increase and/or product concentrations decrease (since these are the only variables in the equation). However, an exogenous process is needed to supply substrates and absorb products so that the cell can maintain a state of balanced nonequilibrium purely through regulation.

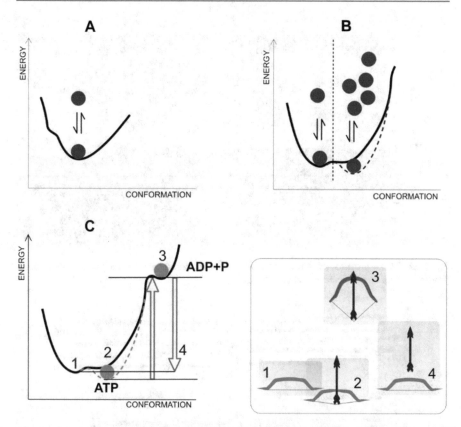

**Fig. 2.22** Structural rearrangement of proteins as a result of energy expenditure: (**a**) non-allosteric protein (one energy minimum), (**b**) allosteric protein (two overlapping energy minima), and (**c**) motor protein (two distinct energy minima) (symbolic presentation). Muscle contraction phases and their associated energy levels are indicated. Spheres mark the role of ligands in stabilizing protein configurations. The bow diagram represents the phase-like properties of muscle contraction

Being part of an organism, the cell is usually provided with reaction substrates and may expel unneeded products. The energy debt incurred in this process is covered by other types of cells. For instance, high levels of glucose required for its spontaneous absorption by muscle cells (where it can power glycolysis) are maintained by hepatic metabolism. Figure 2.26 depicts the spontaneity of glycolysis as a function of $\Delta G^0$, compared with observed values of $\Delta G$ in blood cells.

As can be seen, many stages of glycolysis are not inherently spontaneous (in terms of $\Delta G^0$). Empirically determined $\Delta G$ values indicate that cells maintain their spontaneity by exploiting both direct (phosphorylation) and indirect links to energy sources. Limiting major changes in $\Delta G$ to stages where direct coupling comes into play (Fig. 2.26 a–c) means that these stages are effective in control of the entire process. $\Delta G^0$ values observed at each stage indicate that the process can

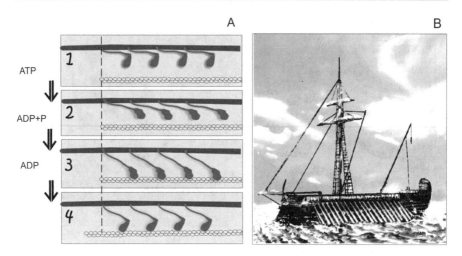

**Fig. 2.23** Motor proteins: (**a**) action of motor proteins in muscle and (**b**) analogies between motor proteins and an oar-powered vessel (Roman galley)

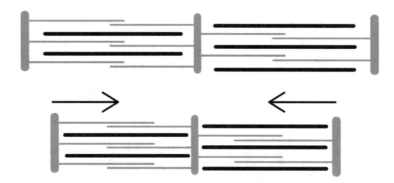

**Fig. 2.24** Telescopic motion as the principle of contraction in motor proteins

easily be reversed (phosphoenolpyruvate synthesis) provided that ATP molecules are not resynthesized and only it is the threshold to be crossed.

Indirect coupling may also benefit processes associated with low-energy structural reconfiguration of proteins (mostly allosteric ones). Such processes are usually powered by changes in ligand concentrations and not by direct links to exergonic reactions. Examples include hemoglobin, receptors, regulatory enzymes, etc.

In general, high-energy reconfigurations exploit direct coupling mechanisms, while indirect coupling is more typical of low-energy processes (however, an important exception to this rule is the coupling of water synthesis to ATP synthesis effected by the hydrogen ion gradient).

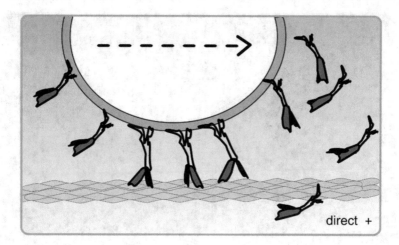

**Fig. 2.25** Kinesin in motion

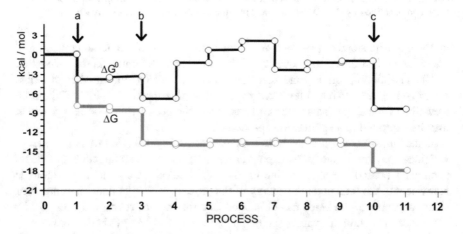

**Fig. 2.26** Spontaneity of glycolysis as a function of $\Delta G^0$ and $\Delta G$. Arrows indicate stages directly coupled to energy sources (stages a, b, and c catalyzed, respectively, by hexokinase, phosphofructokinase-1, and pyruvate kinase)

## 2.6 Energy Conversion Efficiency in Biological Processes

It is impossible to exploit all the energy released by a source. Any system characterized by 100% efficiency would effectively become a *perpetuum mobile*.

We may therefore ask why lossless exploitation of energy is undesirable. The answer lies in the fundamental contradiction between high conversion efficiency and the quantity of conserved energy. Let us again refer to the example of a skier going downhill except this time there are many skiers and the aim is to propel oneself as far

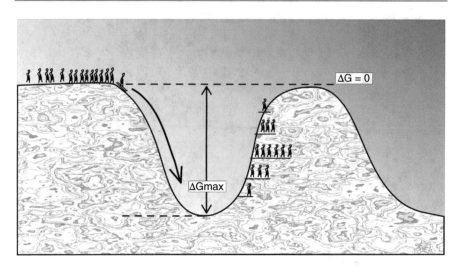

**Fig. 2.27** Reaction efficiency depicted as the distribution of skiers reaching various points on a slope and quantified by the number of skiers multiplied by the elevations they reach

up the opposite slope as possible. While analyzing the outcome of the competition, we will note that the relative density of skiers increases up until a certain elevation and begins to drop off beyond that point. Peak efficiency (the number of skiers multiplied by the attained elevation) occurs at approximately 30–40% of the total elevation of the slope. This value corresponds to the peak fraction of energy which may be exploited in a spontaneous process.

Once the competition is over, each skier is again in possession of stored energy which can pay for further, spontaneous downhill descent. The total number of skiers positioned behind the starting line on the initial slope (at a certain elevation) is equivalent to the total potential energy of the process, while the final distribution of skiers on the opposite slope can be treated as a measure of conserved energy. The difference between both values (potential and conserved energy) is the free energy ($\Delta G$) dissipated during the process while ensuring its spontaneity. Assuming a statistical distribution of skier skill, we can predict that the amount of conserved energy will increase up until a certain elevation (which can be reached by moderately skilled skiers) and then begin to drop off, becoming equal to 0 at an elevation which no skier is able to reach. The elevation reached by the best skier is still lower than that of the starting line; otherwise we would be dealing with a *perpetuum mobile*, i.e., a process which can generate energy despite its $\Delta G$ being 0 (Fig. 2.27).

The efficiency of biological processes is usually below 40%. Synthesizing 1 mol of water yields 56.7 kcal of energy yet can only generate 3.0 mol (2.5 according to some studies) of ATP yielding 7.3 kcal/mol each upon hydrolysis. Thus, the total amount of conserved energy is not higher than 21.9 kcal/mol, which corresponds to an efficiency of 37.4%. It should, however, be remembered that this low efficiency is the price paid for spontaneity, as more than 60% of the energy released by the source is dissipated. A typical human transforms 3 mol (approximately 1.5 kg) of ATP

(ATP $\leftrightarrow$ ADP) in 1 h. Strenuous physical exertion may increase this demand by a factor of 10, though the process itself is dynamic and, on average, no more than 0.1 mol of ATP exists in the organism at any given time.

## 2.7   Entropic Effects

The spontaneity of biological processes is a consequence of enthalpic and entropic changes:

$$\Delta G = \Delta H - T * \Delta S$$

In most cases both phenomena have a measurable effect on spontaneity; however in some situations one clearly dominates the other. In processes where covalent bonds are formed or broken, enthalpy is usually more important than entropy, whereas synthesis of protein complexes and many other noncovalently stabilized structures (such as cellular membranes) relies primarily on entropic effects and is often powered by thermodynamically disadvantageous rearrangement of water molecules, emerging as a result of contact with water-repellent hydrophobic surfaces of compounds introduced into the aqueous environment. Such compounds include polymer chains or particulate organic structures which contain apolar and polar moieties and are therefore water-soluble. In the latter case, aggregation occurs as a result of hydrophobic interactions.

Hydrophobic interaction is not, strictly speaking, a chemical bond—instead, it can be treated as a physical phenomenon related to the thermodynamically disadvantageous interaction between hydrophobic structures and water. Such interaction introduces an entropic force which tries to destroy any structural ordering of water resulting from contact with hydrophobic residues (Fig. 2.28). Aggregation is a means by which such thermodynamically undesirable changes (caused by hydrophobic compounds intruding into the aqueous environment) are reversed. Coupling hydrophobic surfaces to each other, and stabilizing this aggregation with hydrophobic bonds (usually van der Waals bonds) lessens their exposure to water and is therefore thermodynamically preferable.

Synthesis applies to structures which are partly apolar but also contain polar elements. It is necessary for such polar elements to be sufficiently large to dissolve, thus pulling and exposing its apolar groups to water and—as a consequence—forcing aggregation. This mechanism is one of the most basic forms of self-organization.

A classic example of self-association at work is the clustering of phospholipids in the cellular membrane, which determines its tertiary structure (Fig. 2.29). Polypeptide chain folding is considered to be of fundamental importance to this process, although our knowledge of its mechanisms is still somewhat limited. We know that in order for a tertiary structure to emerge, the polypeptide chain immersed in the aqueous environment must exhibit a noneven distribution of polarity, corresponding to nonrandom ordering of hydrophobic amino acids. The apolar fragments of the

**Fig. 2.28** Self-organization powered by hydrophobic interactions (aggregates of apolar structures in an aqueous environment). (**a**) The thermodynamically disadvantageous effect of introducing a hydrophobic compound ($\Delta G > 0.0$) into an aqueous environment and the corresponding effect observed for a polar compound ($\Delta G < 0.0$). (**b**) Covalent (permanent) binding of both compounds results in their selective, nonrandom aggregation and ordering in the presence of water. Arrows indicate salvation tendency

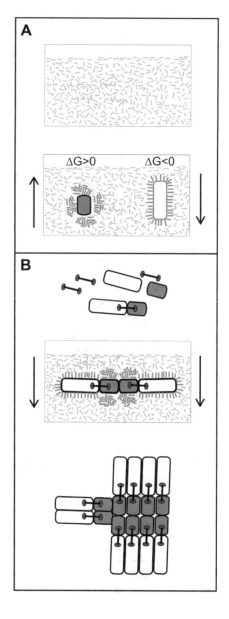

chain undergo aggregation, forming a hydrophobic core which, in turn, is encapsulated by polar fragments. Such thermodynamically optimal folding of the polypeptide chain determines the so-called native structure of the protein. The process is assisted by a special class of proteins called chaperones, and its nature constitutes one of the most important unresolved problems in modern biology. A fully folded globular protein resembles a tightly packed solid, although it usually includes an active pocket, which can be interpreted as a point of access to its apolar

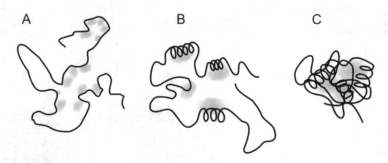

**Fig. 2.29** Globular protein synthesis divided into phases: (**a**) Synthesis of an unraveled polypeptide chain, (**b**) creation of subdomains, and (**c**) structural collapse. Hydrophobic areas marked in color (schematic view)

core. The existence of an active pocket is a prerequisite of forming protein-ligand complexes in aqueous environments.

## 2.8   Energy Requirements of Organisms

Maintaining baseline biological processes requires a supply of energy. In humans, the so-called basal metabolic rate (BMR) is approximately 1500–2000 kcal/day. Normal activity introduces additional demands; thus most humans burn approximately 2000–2500 kcal/day, although strenuous physical exercise may increase this amount to 5000–10,000 kcal/day or even more. Energy demands are normally met through consumption of food. Among nutrients the highest calorie content is found in lipids (9.4 kcal/g). Sugars and proteins are somewhat less energetic, yielding 4.2 kcal/g and 4.3 kcal/g, respectively.

The following list illustrates the approximate calorie content of common foodstuffs:

| | |
|---|---|
| Vegetables (100 g serving) | 20–30 kcal |
| Potatoes (80 g serving) | 60–80 kcal |
| Sugar (1 teaspoon) | 30–40 kcal |
| Egg (domestic chicken) | 100 kcal |
| Veal cutlet | 200–250 kcal |
| Sausage (100 g serving) | 250–300 kcal |
| Butter (100 g) | 750 kcal |
| Vegetable oil (100 g) | 900 kcal |

The average daily intake of carbohydrates (a major component of human diet) is approximately 250 g, which corresponds to 1000 kcal. If we assume that 1 mole of glucose (180 g) affords 38 mol of ATP, 250 g of glucose can cover the cost of synthesizing 52.8 mol of ATP, i.e., 385.3 kcal at an efficiency of 38.5%. This value matches stoichiometric studies of ATP/water synthesis, where 56.7 kcal of energy

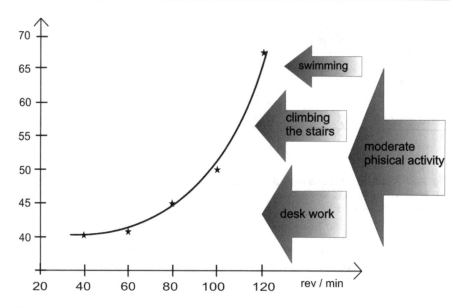

**Fig. 2.30** Relationship between energy expenditure and physical exertion. The diagram illustrates the energy cost incurred by a cyclist pedaling at a rate of 40, 60, 80, 100, and 120 revolutions/min over a period of 6 min against a constant drag force. Measurements are based on the intake of oxygen required for the synthesis of water (Zoladz et al. 1998). Arrows indicate average energy requirements for specific types of activity, while their width denotes the possible range of values

released in the oxidation of hydrogen covers the cost of synthesizing three anhydrous ATP bonds which can subsequently be hydrolyzed for an energy gain of 21.9 kcal. Our calculation therefore assumes that synthesis of 1 mole of water affords 3 mol of ATP.

The energy of anhydrous bonds may be tapped to power endoergic processes. Muscle action requires a major expenditure of energy. There is a nonlinear dependence between the degree of physical exertion and the corresponding energy requirements. Figure 2.30 presents the energy expenditure associated with physical exertion as a function of cyclist pedaling at different rates.

Training may improve the power and endurance of the muscle tissue. Muscle fibers subjected to regular exertion may improve their glycogen storage capacity, ATP production rate, oxidative metabolism, and the use of fatty acids as fuel.

## Suggested Reading

An H, Ordureau A, Körner M, Paulo JA, Harper JW (2020) Systematic quantitative analysis of ribosome inventory during nutrient stress. Nature. 583(7815):303–309. https://doi.org/10.1038/s41586-020-2446-y

Anastasiou D, Poulogiannis G, Asara JM, Boxer MB, Jiang JK, Shen M, Bellinger G, Sasaki AT, Locasale JW, Auld DS, Thomas CJ, Vander Heiden MG, Cantley LC (2011) Inhibition of pyruvate kinase M2 by reactive oxygen species contributes to cellular antioxidant responses. Science. 334(6060):1278–1283. https://doi.org/10.1126/science.1211485

Bazopoulou D, Knoefler D, Zheng Y, Ulrich K, Oleson BJ, Xie L, Kim M, Kaufmann A, Lee Y-T, Dou Y, Chen Y, Quan S, Jakob U (2019) Developmental ROS individualizes organismal stress resistance and lifespan. Nature 576(7786):301–305. https://doi.org/10.1038/s41586-019-1814-y

Bernheim-Groswasser A, Wiesner S, Golsteyn RM, Carlier M-F, Sykes C (2002) The dynamics of actin-based motility depend on surface parameters. Nature 417:308–311

Bertholet AM, Natale AM, Bisignano P, Suzuki J, Fedorenko A, Hamilton J, Brustovetsky T, Kazak L, Garrity R, Chouchani ET, Brustovetsky N, Grabe M, Kirichok Y (2022) Mitochondrial uncouplers induce proton leak by activating AAC and UCP1. Nature 606(7912):180–187. https://doi.org/10.1038/s41586-022-04747-5

Blättler CL, Claire MW, Prave AR, Kirsimäe K, Higgins JA, Medvedev PV, Romashkin AE, Rychanchik DV, Zerkle AL, Paiste K, Kreitsmann T, Millar IL, Hayles JA, Bao H, Turchyn AV, Warke MR, Lepland A (2018) Two-billion-year-old evaporites capture Earth's great oxidation. Science 360(6386):320–323. https://doi.org/10.1126/science.aar2687

Britt RD, Marchiori DA (2019) Photosystem II, poised for $O_2$ formation. Science 366(6463): 305–306. https://doi.org/10.1126/science.aaz4522

Chan CE, Odde DJ (2008) Traction dynamics of filopodia on compliant substrates. Nature 322: 1687–1695

Dibble CC, Barritt SA, Perry GE, Lien EC, Geck RC, DuBois-Coyne SE, Bartee D, Zengeya TT, Cohen EB, Yuan M, Hopkins BD, Meier JL, Clohessy JG, Asara JM, Cantley LC, Toker A (2022) PI3K drives the de novo synthesis of coenzyme A from vitamin B5. Nature. 608(7921): 192–198. https://doi.org/10.1038/s41586-022-04984-8

Domínguez-Martín MA, Sauer PV, Kirst H, Sutter M, Bína D, Greber BJ, Nogales E, Polívka T, Kerfeld CA (2022) Structures of a phycobilisome in light-harvesting and photoprotected states. Nature. 609(7928):835–845. https://doi.org/10.1038/s41586-022-05156-4

Frauenfelder H, Sligar SG, Wolynes PG (1991) The energy landscape and motions of proteins. Science 254:1598–1603

Frazier AE, Thorburn DR, Compton AG (2019) Mitochondrial energy generation disorders: genes, mechanisms, and clues to pathology. J Biol Chem. 294(14):5386–5395. https://doi.org/10.1074/jbc.R117.809194

Garret RH, Grisham CM (1995) Biochemistry. Saunders College Publishing Harcourt Brace College Publishers, New York

Glaser R (2002) Biophysics. Springer

Gottlieb RA, Bernstein D (2015) Metabolism. Mitochondria shape cardiac metabolism. Science 350(6265):1162–1163. https://doi.org/10.1126/science.aad8222

Grant CR, Amor M, Trujillo HA, Krishnapura S, Iavarone AT, Komeili A (2022) Distinct gene clusters drive formation of ferrosome organelles in bacteria. Nature 606(7912):160–164. https://doi.org/10.1038/s41586-022-04741-x

Hardie DG (2011) How cells sense energy. Nature 472:176–177

Van der Heiden MG, Cantley LC, Thompson CB (2009) Understanding the Warburg effect: the metabolic requirements of cell proliferation. Science. 324(5930):1029–1033. https://doi.org/10.1126/science.1160809

Hui S, Ghergurovich JM, Morscher RJ, Jang C, Teng X, Lu W, Esparza LA, Reya T, Zhan L, Yanxiang Guo J, White E, Rabinowitz JD (2017) Glucose feeds the TCA cycle via circulating. Nature. 551(7678):115–118. https://doi.org/10.1038/nature24057

Jang C, Hui S, Lu W, Cowan AJ, Morscher RJ, Lee G, Liu W, Tesz GJ, Birnbaum MJ, Rabinowitz JD (2018) The small intestine converts dietary fructose into glucose and organic acids. Cell Metab. 27(2):351–361.e3. https://doi.org/10.1016/j.cmet.2017.12.016

Junge W, Lill H, Engelbrecht S (1997) ATP-synthase: an electrochemical transducer with rotatory mechanics. TIBS 22:420–423

Kerner J, Hoppel C (2000) Fatty acid import into mitochondria. Biochim Biophys Acta. 1486(1):
     1–17. https://doi.org/10.1016/s1388-1981(00)00044-5
Knoefler D, Thamsen M, Koniczek M, Niemuth NJ, Diederich AK, Jakob U (2012) Quantitative
     in vivo redox sensors uncover oxidative stress as an early event in life. Mol Cell. 47(5):767–776.
     https://doi.org/10.1016/j.molcel.2012.06.016
Kornberg MD, Bhargava P, Kim PM, Putluri V, Snowman AM, Putluri N, Calabresi PA, Snyder
     SH (2018) Dimethyl fumarate targets GAPDH and aerobic glycolysis to modulate immunity.
     Science 360(6387):449–453. https://doi.org/10.1126/science.aan4665
Kravchuk V, Petrova O, Kampjut D, Wojciechowska-Bason A, Breese Z, Sazanov L (2022) A
     universal coupling mechanism of respiratory complex I. Nature. 609(7928):808–814. https://
     doi.org/10.1038/s41586-022-05199-7
Lane N, Martin W (2010) The energetics of genome complexity. Nature 467(7318):929–934.
     https://doi.org/10.1038/nature09486
Langford GM (2002) Myosin-V, a versatile motor for short-range vesicle transport. Traffic 3:859–
     865
Lin J, Nicastro D (2018) Asymmetric distribution and spatial switching of dynein activity generates
     ciliary motility. Science 360(6387). https://doi.org/10.1126/science.aar1968
Lodish H, Berk A, Zipursky SL, Matsudaira P, Baltimore D, Darnell J (1999) Molecular cell
     biology. W. H. Freeman
Miller M, Craig JW, Drucker WR, Woodward H (1956) The metabolism of fructose in man. Yale J
     Biol Med 29(3):335–360
Mo K, Zhang Y, Dong Z, Yang Y, Ma X, Feringa BL, Zhao D (2022) Intrinsically unidirectional
     chemically fuelled rotary molecular motors. Nature. 609(7926):293–298. https://doi.org/10.
     1038/s41586-022-05033-0
Nathan JA (2022) Metabolite-driven antitumor immunity. Science. 377(6614):1488–1489. https://
     doi.org/10.1126/science.ade3697
Nicholls DG, Ferguson S (2002) Bioenergetics. Academic Press
Palsson-McDermott EM, Curtis AM, Goel G, Lauterbach MA, Sheedy FJ, Gleeson LE, van den
     Bosch MW, Quinn SR, Domingo-Fernandez R, Johnston DG, Jiang JK, Israelsen WJ, Keane J,
     Thomas C, Clish C, Vander Heiden M, Xavier RJ, O'Neill LA (2015) Pyruvate kinase M2
     regulates Hif-1α activity and IL-1β induction and is a critical determinant of the warburg effect
     in LPS-activated macrophages. Cell Metab. 21(1):65–80. https://doi.org/10.1016/j.cmet.2014.
     12.005
Pascal R (2019 May) A possible non-biological reaction framework for metabolic processes on
     early Earth. Nature. 569(7754):47–49. https://doi.org/10.1038/d41586-019-01322-3
Peusner L (1974) Concepts in bioenergetics. Prentice Hall, New Jersey
Pollard TD, Earnshaw W (2002) Cell biology. Saunders, London
Rich P (2003) The cost of living. Nature 421:583–584
Roca FJ, Whitworth LJ, Prag HA, Murphy MP, Ramakrishnan L (2022) Tumor necrosis factor
     induces pathogenic mitochondrial ROS in tuberculosis through reverse electron transport.
     Science 376(6600):eabh2841. https://doi.org/10.1126/science.abh2841
Salway JG (2004) Metabolism at a glance. Blackwell Publishing
Schulz TJ, Zarse K, Voigt A, Urban N, Birringer M, Ristow M (2007) Glucose restriction extends
     Caenorhabditis elegans life span by inducing mitochondrial respiration and increasing oxidative
     stress. Cell Metab 6(4):280–293. https://doi.org/10.1016/j.cmet.2007.08.011·
Seaman C, Wyss S, Piomelli S (1980) The decline in energetic metabolism with aging of the
     erythrocyte and its relationship to cel death. Am J Hematol 8:31–42
Stryer L (2002) Biochemistry. W. H. Freeman and Company
Sumino Y, Nagai KH, Shitaka Y, Yoshikawa K, Chaté H, Oiwa K (2012) Large-scale vortex lattice
     emerging from collectively moving microtubules. Nature 483:448–452
Taylor SR, Ramsamooj S, Liang RJ, Katti A, Pozovskiy R, Vasan N, Hwang SK, Nahiyaan N,
     Francoeur NJ, Schatoff EM, Johnson JL, Shah MA, Dannenberg AJ, Sebra RP, Dow LE,
     Cantley LC, Rhee KY, Goncalves MD (2021) Dietary fructose improves intestinal cell survival

and nutrient absorption. Nature. 597(7875):263–267. https://doi.org/10.1038/s41586-021-03827-2

Thompson JJ (1969) An introduction to chemical energetics SI Edition Longmans, London

Vicsek T (2012) Swarming microtubules. Nature 483:411–412

Voet D, Voet JG, Pratt CW (1999) Fundamentals of biochemistry. Wiley, New York

Weinert T, Skopintsev P, James D, Dworkowski F, Panepucci E, Kekilli D, Furrer A, Brünle S, Mous S, Ozerov D, Nogly P, Wang M, Standfuss J (2019) Proton uptake mechanism in bacteriorhodopsin captured by serial synchrotron crystallography. Science 365(6448):61–65. https://doi.org/10.1126/science.aaw8634

Yang H, Wang X, Xiong X, Yin Y (2016) Energy metabolism in intestinal epithelial cells during maturation along the crypt-villus axis. Sci Rep 6:31917. https://doi.org/10.1038/srep31917

Zhu J, Schwörer S, Berisa M, Kyung YJ, Ryu KW, Yi J, Jiang X, Cross JR, Thompson CB (2021) Mitochondrial NADP(H) generation is essential for proline biosynthesis. Science. 372(6545): 968–972. https://doi.org/10.1126/science.abd5491

# Information: Its Role and Meaning in Organisms

**3**

*Information determines the function of regulatory mechanisms by reducing the entropy.*

### Abstract

*Information is necessary in regulatory mechanisms* which maintain a steady state of activity in individual cells as well as the whole organism. This state corresponds to a genetically encoded program. Without regulation biological processes would become progressively more and more chaotic. In living cells the primary source of information is genetic material. Studying the role of information in biology involves signaling (i.e., spatial and temporal transfer of information) and storage (preservation of information).

Regarding the role of the genome, we can distinguish three specific aspects of biological processes: steady-state genetics, which ensure cell level and body homeostasis; genetics of development, which controls cell differentiation and organo-genesis; and evolutionary genetics, which drives speciation. A systemic approach to these phenomena must account for the quantitative and qualitative properties of information, explaining that the former are associated with receptor proteins, while the latter correspond to biological effectors.

The ever-growing demand for information, coupled with limited storage capacities, has resulted in a number of strategies for minimizing the quantity of the encoded information that must be preserved by living cells. In addition to combinatorial approaches based on noncontiguous gene structure, self-organization plays an important role in cellular machinery. Nonspecific interactions with the environment give rise to coherent structures despite the lack of any overt information store. These mechanisms, honed by evolution and ubiquitous in living organisms, reduce the need to directly encode large quantities of data by adopting a systemic approach to information management. Our work represents an attempt to employ the similar mechanisms in the teaching process.

© The Author(s) 2023

L. Konieczny et al., *Systems Biology*, https://doi.org/10.1007/978-3-031-31557-2_3

**Keywords**

Need for information · Information and entropy · Source of information ·
Information storage · Steady-state genetics · Genetics of development ·
Evolutionary genetics · Indirect encoding of information · Compartmentalization

1. How can information be quantified, and why is this important?
2. DNA damage—what are the basic repair mechanisms?
3. What is the relation between the 3D structure of genetic code and its function?
4. Discontinuity of genes—its role in biology.
5. What is the reason for the unidirectionality ($5 \rightarrow 3$) of DNA synthesis?
6. Epigenetics—the mechanism and its biological role.
7. Extragenetic storage of information in the cell.
8. What processes determine the variability and diversity of embryonal development? What about evolutionary development?
9. When does the "multiple attempts" (k) requirement become relevant in trying to reach a goal?
10. What enables the possibility of targeted action through the "p" route?
11. Detoxification—theoretical premise and mechanism.
12. The red blood cell does not have a nucleus, but is capable of function—where, then, is the necessary information stored?
13. What basic information is required for an organism to be formed from autonomous cells?
14. How are processes which consist of multiple stages encoded?
15. In general, how does biology solve problems which carry a high degree of uncertainty?

The mechanisms described in this chapter can be experimented with using two web applications we provide for the reader's convenience: The Information Probability (IP) tool, available at https://ip.sano.science, which simulates the likelihood of achieving the given goal depending on the number of repetitions and elementary probability.

## 3.1    Information as a Quantitative Concept

The science of regulatory mechanisms has recently emerged from the shadow of structural and functional research. Observational evidence indicates that most of the information encoded in the genome serves regulatory purposes; hence scientific interest in biological regulation is growing rapidly.

**Fig. 3.1** Road forks with varying numbers of exits

Our emerging knowledge of regulatory mechanisms calls for a quantitative means of describing information. The twentieth century and the 1940s, in particular, have brought about significant progress in this matter, owing to the discoveries by N. Wiener, G. Walter, W. R. Ashby, C. E. Shannon, and others. In order to explain the nature of information and the role it plays in biological systems, we first need to rehash our terminology.

Information is commonly understood as a transferable description of an event or object. Information transfer can be either spatial (communication, messaging, or signaling) or temporal (implying storage).

From a quantitative point of view, information is not directly related to the content of any particular message, but rather to the ability to make an informed choice between two or more possibilities. Thus, information is always discussed in the context of some regulated activity where the need for selection emerges.

Probability is a fundamental concept in the theory of information. It can be defined as a measure of the statistical likelihood that some event will occur.

If the selection of each element from a given set is equally probable, then the following equation applies:

$$p = 1/N$$

where $N$ is the number of elements in the set.

Let us consider a car driving down a straight road. If the layout of the road does not force the driver to make choices, the information content of the driving process is nil. However, when forks appear and a decision has to be made, the probability of choosing the correct exit is equal to $p = 1/2$, $p = 1/3$, or $p = 1/5$, for a two-, three-, or five-way fork, respectively.

Choosing the correct route by accident becomes less probable as the number of exits increases (Fig. 3.1).

If more than one exit leads to our intended destination, then the associated probability increases, becoming equal to 2/3, 3/5, or 4/5 for a five-way fork where two, three, or four exits result in the correct direction of travel. Clearly, if all exits are good, the probability of making a correct choice is given as

$$p = 5/5 = 1.0$$

The value 1.0 implies certainty: no matter which exit we choose, we are sure to reach our destination.

Probabilities are additive and multiplicative. Total probability is a sum of individual probabilities whenever an alternative is involved, i.e., when we are forced to choose one solution from among many, provided that some of the potential choices are correct and some are wrong. If a four-way fork includes two exits which lead to our destination, the probability of accidentally making the right decision (i.e., choosing one of the two correct exits) is equal to 1/2, according to the following formula:

$$p = p_1 + p_2 = 1/4 + 1/4 = 1/2$$

which means that, given four exits, the likelihood of choosing an exit that leads to our destination is the sum of the individual probabilities of choosing any of the correct exits (1/4 in each case).

If, in addition to the fork mentioned above, our route includes an additional five-way fork with just one correct exit, we can only reach our destination if we make correct choices on both occasions (this is called a conjunction of events). In such cases, probabilities are multiplicative. Thus, the probability of choosing the right route is given as

$$p = (p_1 + p_2)^* p_3 = (1/4 + 1/4)^* \, 1/5 = 1/10$$

The larger the set of choices, the lower the likelihood making the correct choice by accident and—correspondingly—the more information is needed to choose correctly. We can therefore state that an increase in the cardinality of a set (the number of its elements) corresponds to an increase in selection indeterminacy. This indeterminacy can be understood as a measure of "*a priori* ignorance." If we do not know which route leads to the target and the likelihood of choosing each of the available routes is equal, then our a priori ignorance reaches its maximum possible value.

The difficulty of making the right choice depends not only on the number of potential choices but also on the conditions under which a choice has to be made. This can be illustrated by a lottery where 4 of 32 numbers need to be picked.

The probability of making one correct selection is 4/32. The probability that two numbers will be selected correctly can be calculated as a product of two distinct probabilities and is equal to

$$p_2 = (4/32)^* \, (3/31)$$

Similarly, the probability that all of our guesses will be correct is given as
$p_4 = (4/32)^*(3/31) * (2/30) * (1/29)$, which is equal to $p_{WIN} = 0.0000278$.

The above value denotes our chances for winning the lottery. In contrast, the corresponding probability of a total loss (i.e., not selecting any of the four lucky numbers) is $p_{LOSS} = 0.5698$. If someone informs us that we have lost and that not a single one of our selections was correct, we should not be surprised, as this is a fairly likely outcome. However, a message telling us that we have won carries a far higher *information content*—not due to any emotional considerations but because the odds of winning are extremely low.

In 1928 R. V. L. Hartley defined information quantity ($I$) as

$$I = -\log_2(p) \, [\text{bit}]$$

(For a base-2 logarithm, the result is given in bits.)

Referring to the above example, the information quantity contained in a message indicating that we have won the lottery is given as

$$I_{\text{WIN}} = -\log_2 0.0000278 = 15.2 \, \text{bit}$$

A corresponding message informing us of a total loss carries significantly less information:

$$I_{\text{LOSS}} = -\log_2 0.5698 = 0.812 \, \text{bit}$$

The bit is a basic unit of information, corresponding to the quantity of information required to make a choice between two equally probable events or objects.

In the above example, we focused on two unambiguous scenarios, i.e., hitting the jackpot (all selections correct) or losing entirely (all selections incorrect). If you wish to learn more, consider the following example.

Arriving at a probabilistic measure of entropy requires us to consider all possible outcomes (called realizations). This includes partial wins. For example, the probability of getting *exactly* two numbers right is

$$
\begin{aligned}
p_{(2)} ={}& 4/32^*3/31^*28/30^*27/29 \\
&+4/32^*28/31^*3/30^*27/29 \\
&+4/32^*28/31^*27/30^*3/29 \\
&+28/32^*4/31^*3/30^*27/29 \\
&+28/32^*27/31^*4/30^*3/29 \\
&+28/32^*4/31^*27/30^*3/29
\end{aligned}
$$

This expression corresponds to the likelihood of arriving at the same end result (two correct guesses) in various ways (C, correct guess; W, wrong guess):

$$\text{CCWW} + \text{CWCW} + \text{CWWC} + \text{WCCW} + \text{WWCC} + \text{WCWC}$$

As the goal can be reached in six different ways and the probability of each sequence is equal, we may calculate the value of $p_{(2)}$ using a simplified formula:

$$p_{(2)} = 6^*[(3^*4^*27^*28)/(32^*31^*30^*29)]$$

In the case of our lottery which has five different outcomes (A, four correct guesses; B, three correct guesses; C, two correct guesses; D, one correct guess;

and E, no correct guesses), the measure of uncertainty is the average quantity of information involved in making a selection.

C. E. Shannon was the first to relate statistical uncertainty to physical entropy, arriving at the formula:

$$H = - \Sigma p_i \log_2(p_i)$$

where $H$ is the information entropy and $n$ is the number of possible outcomes.

Determining $H$ has practical consequences as it enables us to compare different, seemingly unrelated situations.

$H$ indicates the (weighted) average quantity of information associated with the realization of an event for which the sum of all $p_i$ equals 1.

The mathematical formula for $H$ in the case of the presented lottery is

$$H = - 0.36432^* \log_2(0.36432) - 0.06306^* \log_2(0.06306) - 0.00311^* \log_2(0.00311)$$
$$- 0.0000278^* \log_2(0.0000278) - 0.5698^* \log_2(0.5698) = 1.271$$

The first component represents the "one correct guess" outcome; the second, "two correct guesses"; and so on until the final component where none of the selected numbers are correct. (Notice the sum of probabilities of all cases is equal to 1.)

Entropy determines the uncertainty inherent in a given system and therefore represents the relative difficulty of making the correct choice. For a set of possible events, it reaches its maximum value if the relative probabilities of each event are equal. Any information input reduces entropy—we can therefore say that changes in entropy are a quantitative measure of information. This can be denoted in the following way:

$$I = H_i - H_r$$

where $H_i$ indicates the initial entropy and $H_r$ stands for the resulting entropy.

Although physical and information entropy are mathematically equivalent, they are often expressed in different units. Physical entropy is usually given in J/(mol*degree), while the standard unit of information entropy is 1 bit.

Physical entropy is highest in a state of equilibrium, i.e., lack of spontaneity ($\Delta G = 0, 0$), which effectively terminates the given reaction. Regulatory processes which counteract the tendency of physical systems to reach equilibrium must therefore oppose increases in entropy. It can be said that a steady inflow of information is a prerequisite of continued function in any organism.

As selections are typically made at the entry point of a regulatory process, the concept of entropy may also be applied to information sources. This approach is useful in explaining the structure of regulatory systems which must be "designed" in a specific way, reducing uncertainty and enabling accurate, error-free decisions.

One of the models which can be used to better illustrate this process is the behavior of social insects which cooperatively seek out sources of food (Fig. 3.2).

Nonrandom pathing is a result of the availability of information, expressed, e.g., in the span of the arc (1/8, 1/16, or 1/32 of the circumference of a circle).

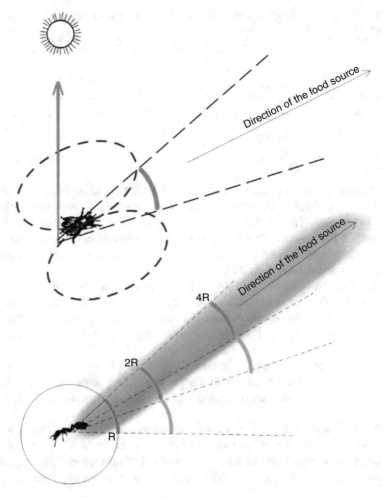

**Fig. 3.2**  Information regarding the location of a food source conveyed using visual (bee dance) or olfactory (ant pheromones—dark strip) cues

The ability to interpret directional information presented by a bee that has located a source of food means that other bees are not forced to make random decisions. Upon returning to the hive, the bee performs a "dance," where individual movements indicate the approximate path to the food source.

If no directional information is available, the dance is random as the source may lie anywhere in relation to the hive. However, if there is specific information, the bee traverses an arc, where the width (in relation to the full circumference of a circle) corresponds to the quantity of information. For instance, if the bee traverses 1/16 of the circumference of a circle (22.5°), the quantity of the conveyed information is 4 bits. Corrections can be introduced by widening the radius of the circle, along the way to the target.

**Table 3.1** Comparison of the probability of reaching a given goal and the amount of required information relative to the distance from the starting point

| Distance | Number of choices | | Likelihood of making the correct choice | | Information quantity (bits) | |
|---|---|---|---|---|---|---|
| | Per stage | Total | Per stage | Total | Per stage | Total |
| 1R (2.5 cm) | 8 | 8 | 1/8 | 1/8 | 3 | 3 |
| 2 R (5.0 cm) | 2 | 8*2 = 16 | 1/2 | 1/8*1/2 | 1 | 3 + 1 = 4 |
| 4 R (10.0 cm) | 2 | 8*2*2 = 32 | 1/2 | 1/8*1/2*1/2 | 1 | 3 + 1 + 1 = 5 |

The fire ant exudes a pheromone which enables it to mark sources of food and trace its own path back to the colony. In this way, the ant conveys pathing information to other ants. The intensity of the chemical signal is proportional to the abundance of the source. Other ants can sense the pheromone from a distance of several (up to a dozen) centimeters and thus locate the source themselves. Figure 3.2 and Table 3.1 present the demand of information relative to the distance at which information can be detected by an insect, for a given length of its path.

The quantity of information required to locate the path at a distance of 2.5 cm is 3 bits. However, as the distance from the starting point increases and the path becomes more difficult to follow, the corresponding demand for information also grows.

As can be expected, an increase in the entropy of the information source (i.e., the measure of ignorance) results in further development of regulatory systems—in this case, receptors capable of receiving signals and processing them to enable accurate decisions.

Over time, the evolution of regulatory mechanisms increases their performance and precision. The purpose of various structures involved in such mechanisms can be explained on the grounds of information theory. The primary goal is to select the correct input signal, preserve its content, and avoid or eliminate any errors.

## 3.2    Reliability of Information Sources

An information source can be defined as a set of messages which assist the recipient in making choices. However, in order for a message to be treated as an information source, it must first be read and decoded.

An important source of information is memory which can be further divided into acquired memory and genetic (evolutionary) memory.

Acquired memory is the set of messages gathered in the course of an individual life. This memory is stored in the nervous system and—to some extent—in the immune system, both capable of remembering events and amassing experience.

However, a more basic source of information in the living world is genetic memory. This type of memory is based upon three dissimilar (though complementary) channels, which differ with respect to their goals and properties:

1. Steady-state genetics, including the "software" required for normal functioning of mature cells and the organism as a whole. This type of information enables biological systems to maintain homeostasis.
2. Development genetics, which guides cell differentiation and the development of the organism as a whole (also called epigenetics).
3. Evolutionary genetics, including mechanisms which facilitate evolutionary progress.

Figure 3.3 presents a simplified model of the genome—a single chromosome with all three channels indicated (dark bands represent the DNA available for transcription of genetic material).

Dark bands represent conventionally the total volume of DNA, i.e., the DNA available for transcription in a mature specialized cell.

The role of the genome is to encode and transfer information required for the synthesis of self-organizing structures in accordance with evolutionary programming, thereby enabling biological functions. Information transfer can be primary (as in the synthesis of RNA and directly used proteins) or secondary (as observed in the synthesis of other structures which ensure cell homeostasis).

## 3.2.1  Steady-State Genetics

Genetic information stored in nucleotide sequences can be expressed and transmitted in two ways:

A.  Via replication (in cell division)
B.  Via transcription and translation (also called gene expression—enabling cells and organisms to maintain their functionality; see Fig. 3.4).

Both processes act as effectors and can be triggered by certain biological signals transferred on request.

Gene expression can be defined as a sequence of events which lead to the synthesis of proteins or their products required for a particular function. In cell division, the goal of this process is to generate a copy of the entire genetic code (S phase), whereas in gene expression only selected fragments of DNA (those involved in the requested function) are transcribed and translated. Reply to the trigger comes in the form of synthesis (and thus activation) of a specific protein. Information is transmitted via complementary nucleic acid interactions, DNA–DNA, DNA–RNA, and RNA–RNA, as well as interactions between nucleic acid chains and proteins (translation). Transcription calls for exposing a section of the cell's genetic code, and although its product (RNA) is short-lived, it can be recreated on demand, just like a carbon copy of a printed text. On the other hand, replication affects the entire genetic material contained in the cell and must conform to stringent precision requirements, particularly as the size of the genome increases.

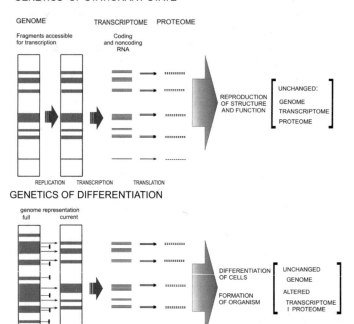

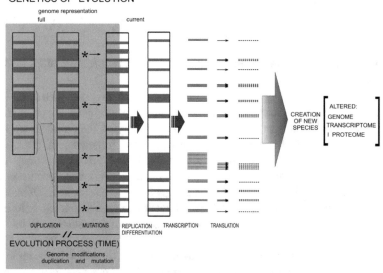

**Fig. 3.3** Simplified view of the genome (bar) and its basic functions

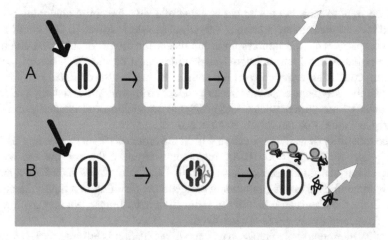

**Fig. 3.4** Simplified diagram of replication (**a**) and transcription/translation (**b**) processes. Arrows indicate the flow of regulatory signals which control syntheses (see Chap. 4)

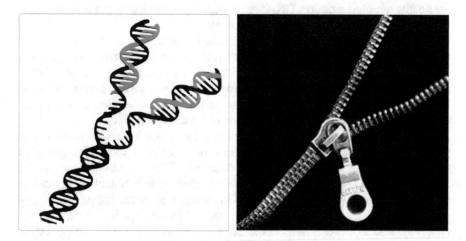

**Fig. 3.5** Similarities between unfastening a zipper and uncoiling the DNA helix in the process of replication

## 3.2.2  Replication and Its Reliability

The magnitude of effort involved in replication of genetic code can be visualized by comparing the DNA chain to a zipper (Fig. 3.5). Assuming that the zipper consists of three pairs of interlocking teeth per centimeter (300 per meter) and that the human genome is made up of 3 billion ($3*10^9$) base pairs, the total length of our uncoiled DNA in "zipper form" would be equal to $1*10^4$ km or 10,000 km—roughly twice the distance between Warsaw and New York.

If we were to unfasten the zipper at a rate of 1 m/s, the entire unzipping process would take approximately 3 months—the time needed to travel 10,000 km at 1 m/s. This comparison should impress upon the reader the length of the DNA chain and the precision with which individual nucleotides must be picked to ensure that the resulting code is an exact copy of the source. It should also be noted that for each base pair the polymerase enzyme needs to select an appropriate matching nucleotide from among four types of nucleotides present in the solution and attach it to the chain (clearly, no such problem occurs in zippers).

The reliability of an average enzyme is on the order of $10^3$–$10^4$, meaning that one error occurs for every 1000–10,000 interactions between the enzyme and its substrate. Given this figure, replication of $3*10^9$ base pairs would introduce approximately three million errors (mutations) per genome, resulting in a highly inaccurate copy. Since the observed reliability of replication is far higher, we may assume that some corrective mechanisms are involved.

Really, the remarkable precision of genetic replication is ensured by DNA repair processes and in particular by the corrective properties of polymerase itself. Its enzymatic association with exonuclease acting in the $3' \rightarrow 5'$ direction increases the fidelity of polymerization. DNA repair works by removing incorrect adducts and replacing them with proper nucleotide sequences.

The direction of anti-parallel DNA strands is determined by their terminating nucleotides or, more specifically, by their hydroxyl groups (the sole participants of polymerization processes) attached to $3'$ and $5'$ carbons of deoxyribose. In the $5' \rightarrow 3'$ direction, a free nucleotide ($5'$-triphosphate nucleotide) may attach itself to the $3'$-carbon hydroxyl group, whereas in the opposite direction, only the $5'$ carbon hydroxyl group may be used to extend the chain (Fig. 3.6).

The proofreading properties of polymerase are an indispensable condition of proper replication of genetic material. However, they also affect the replication process itself, by enforcing one specific direction of DNA synthesis, namely, the $5' \rightarrow 3'$ direction (Fig. 3.7). For reasons related to the distribution of energy, this is the only direction in which errors in the DNA strand may be eliminated by cleaving the terminal nucleotide and replacing it with a different unit. In this process, the energy needed to create a new diester bond is carried by the free $5'$-triphosphate nucleotide, whereas in the $3' \rightarrow 5'$ direction, the required energy can only come from the terminating nucleotide which thus cannot be cleaved for repair. Doing so would, however, make it impossible to attach another nucleotide at the end of the chain. This is why both prokaryotic and eukaryotic organisms can only accurately replicate their genetic code in the $5' \rightarrow 3'$ direction. An important consequence of this fact is the observed lack of symmetry in replication of complementary DNA strands (Fig. 3.8).

Linear synthesis is only possible in the case of the $5' \rightarrow 3'$ copy based on the $3' \rightarrow 5'$ template. Synthesis of the $3' \rightarrow 5'$ copy proceeds in a piecemeal fashion, through the so-called Okazaki fragments which are synthesized sequentially as the replication fork progresses and template loops are formed. This looping mechanisms result in the desired $3' \rightarrow 5'$ direction of synthesis (Fig. 3.8).

The polymerase enzyme itself exhibits as necessary both polymerase and exonuclease activities. We should hence note that the availability of a DNA template is not

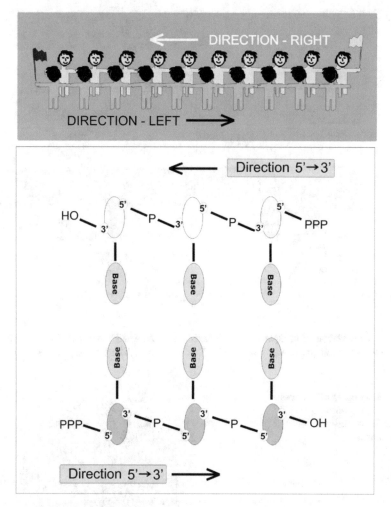

**Fig. 3.6** Anti-parallel arrangement of DNA strands, mandating structural complementarity

sufficient to begin polymerization. In order to attach itself to the chromatid and commence the process, the polymerase enzyme must first interface with a complementary precursor strand, which it can then elongate. This short fragment is called a primer (Fig. 3.9). Its ability to bind to polymerase enzymes has been exploited in many genetic engineering techniques.

The DNA replication fork involves an unbroken, continuous (though unwound) template. Since polymerase cannot directly attach to either of its strands, the primer must first be synthesized by RNA polymerase (i.e., transcriptase). This enzyme does not require a primer to initiate its function and can attach directly to the one strained template.

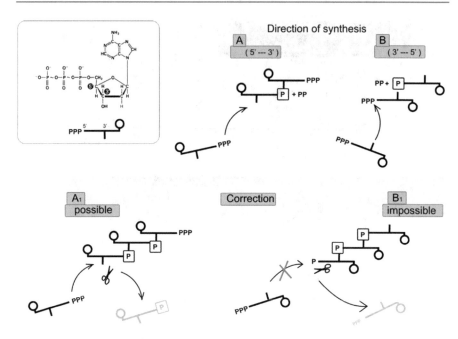

**Fig. 3.7** Polymerization of DNA strands in the $5' \rightarrow 3'$ (**A**) and $3' \rightarrow 5'$ (**B**) directions, together with potential repair mechanisms (A1 and B1, respectively). The inset represents a simplified nucleotide model

**Fig. 3.8** Looping as a means of achieving unidirectional replication of anti-parallel DNA strands

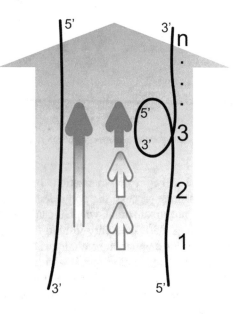

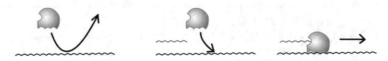

**Fig. 3.9** A nucleic acid primer complementary to the original DNA template is required by polymerase, which also exhibits exonuclease activity in the $3' \rightarrow 5'$ direction

Polymerase carries out DNA synthesis by elongating the RNA fragment supplied to it by RNA polymerase. Therefore, each incidence of DNA synthesis must begin with transcription. The transcriptase enzyme responsible for assisting DNA replication is called the primase.

The synthesized RNA fragments are paired up with complementary nucleotides and attached to the strand which is being elongated by DNA polymerase. They are then replaced by DNA nucleotides, and the complementation to template fragment is ligated. This seemingly complicated process and the highly evolved structure of polymerase itself are necessary for reducing the probability of erroneous transcription.

Permanent changes introduced in the genetic transcription process are called mutations. They can result from random events associated with the function of the genome, or from environmental stimuli, either physical (e.g., UV radiation) or chemical. Mutations which do not compromise the complementarity of DNA often go unrecognized by proofreading mechanisms and are never repaired.

In addition to direct changes in genetic code, errors may also occur as a result of the imperfect nature of information storage mechanisms. Many mutations are caused by the inherent chemical instability of nucleic acids: for example, cytosine may spontaneously convert to uracil. In the human genome, such an event occurs approximately 100 times per day; however uracil is not normally encountered in DNA, and its presence alerts defensive mechanisms which correct the error.

Another type of mutation is spontaneous depurination, which also triggers its own, dedicated error correction procedure. Cells employ a large number of corrective mechanisms—some capable of mending double-strand breaks or even recovering lost information based on the contents of the homologous chromatid (Fig. 3.10). DNA repair mechanisms may be treated as an "immune system" which protects the genome from loss or corruption of genetic information.

The unavoidable mutations which sometimes occur despite the presence of error correction mechanisms can be masked due to doubled presentation (alleles) of genetic information. Thus, most mutations are recessive and not expressed in the phenotype.

As the length of the DNA chain increases, mutations become more probable. It should be noted that the number of nucleotides in DNA is greater than the relative number of amino acids participating in polypeptide chains. This is due to the fact that each amino acid is encoded by exactly three nucleotides—a general principle which applies to all living organisms.

Information theory tells us why, given four available nucleotides, a three-nucleotide codon carries the optimal amount of information required to choose one

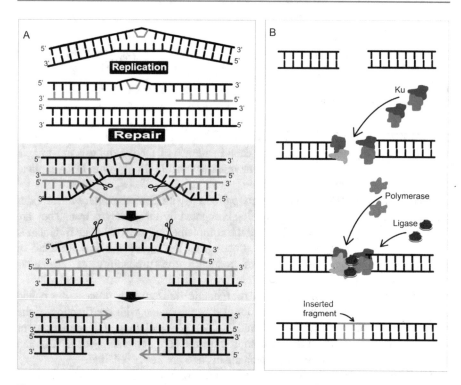

**Fig. 3.10** Repairing double-strand DNA breaks: (**a**) using the sister chromatid to recover missing information and (**b**) reattaching severed strands via specific proteins

of 20 amino acids. The quantity of information carried by three nucleotides, each selected from a set of four, equals $I_3 = -\log_2(1/4^* \ 1/4^* \ 1/4) = 6$ bits, whereas in order to choose one of 20 amino acids, $I_{20} = -\log_2(1/20) = 4.23$ bits of information is required.

If the codon were to consist of two nucleotides, it would carry $I_2 = -\log_2(1/4^*1/4) = 4$ bits of information, which is insufficient to uniquely identify an amino acid. This is why nucleotide triplets are used to encode amino acids, even though their full information potential is not exploited (a nucleotide triplet could theoretically encode $4^*4^*4 = 64$ nucleotides).

If the DNA were to consist of only two base types, the minimum number of nucleotides required to encode 20 amino acids would be 5:

$$I = -\log_2(1/2^*1/2^*1/2^*1/2^*1/2) = -\log_2(1/32) = 5 \text{ bit}$$

In this case, redundancy would be somewhat reduced, but the DNA chain would become far longer, and the likelihood of harmful mutations would increase accordingly.

Considering the reliability of genetic storage mechanisms, the selected encoding method appears optimal. We should, however, note that despite the presence of many safeguards, errors cannot be completely eliminated.

### 3.2.2.1 Telomeres

Telomerase is a peculiar polymerase—an enzyme which elongates DNA strands, enabling reconstruction of the terminal fragments of chromosomes called telomeres. These fragments are truncated at each replication as polymerase cannot carry through to the very end of the DNA molecule due to the specific mechanism by which the lagging strand is synthesized.

Thus, telomerase elongates one of the DNA strands in the $5' \rightarrow 3'$ direction. Acting as a reverse transcriptase, it utilizes its own RNA matrix which binds to the enzyme and consists of repeating AAUCCC fragments (Fig. 3.11). This, in turn, enables attachment of a normal DNA polymerase, in the usual fashion—via an RNA primer—and completion of the chromosome through synthesis of the complementary strand. The need for such a convoluted solution arises due to the anti-parallel

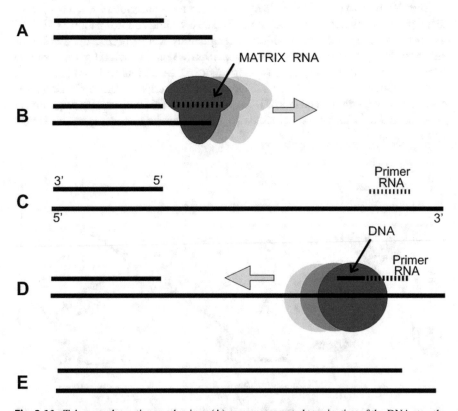

**Fig. 3.11** Telomere elongation mechanism: (**A**) uneven, truncated termination of the DNA strands, (**B**) attachment of telomerase to the leading strand and initiation of synthesis, (**C**) elongated leading strand with an attached primer, (**D**) onset of polymerase activity, and (**E**) completed telomere

arrangement of DNA strands and differences in the way in which each strand is synthesized during replication. Crucially, loss of telomerase activity in differentiated cells and the corresponding limitation in the number of possible divisions are strongly tied to the strategy and philosophy of nature.

### 3.2.2.2 Function-Oriented Organization of the Genome

The control of the intracellular transcription is enabled by the specific, function-related 3D organization of chromatin forming loops comprising promoter region and regulatory fragments including an enhancer required to generate a given biological activity. The location of such loops is determined DNA sequences recognized by the transcription factor called CTCF responsible for anchoring loops at the given locus. The loop itself is formed by protein complex called cohesion, which has the properties of the molecular motor using ATP to migrate along the chain of DNA and create anchor at two initially separated loci. The resulted structures form topologically associating domains called TADs. They emerge when cohesion terminates its activity at the contact with CTCF. This mechanism creates the link between the spatial structure of a chromosome and its biological activity (Fig. 3.12).

The finally formed complex ready to initiate the transcription is composed of many protein constituents assembled to produce coordinated information needed for the proper starting and reading of DNA template by RNA polymerase. Beside the enhancer also other necessary components are enjoined including transcription factors, the protein pulling apart two strands of DNA and as well proteins which allow conquering the resistance arising when transcription proceeds along the protected double-strained chains of DNA by histones. This indicates the requirement of some extra information necessary to make signaling in complex biological systems possible.

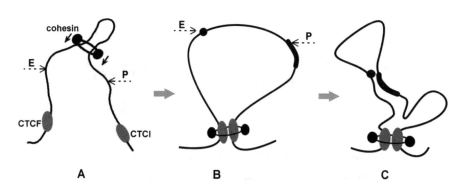

**Fig. 3.12** Organization of chromatin structure. Formation of TADs: (**a**) unlooped form (E and P enhancer and promoter regions, respectively), (**b**) looped form with conventionally marked regulatory and promoter-associating proteins, and (**c**) structure-initiating transcription activity

### 3.2.3 Gene Expression and Its Fidelity

Fidelity is, of course, fundamentally important in DNA replication as any harmful mutations introduced in its course are automatically passed on to all successive generations of cells. In contrast, transcription and translation processes can be more error-prone as their end products are relatively short-lived. Of note is the fact that faulty transcripts appear in relatively low quantities and usually do not affect cell functions, since regulatory processes ensure continued synthesis of the required substances until a suitable level of activity is reached.

Nevertheless, it seems that reliable transcription of genetic material is sufficiently significant for cells to have developed appropriate proofreading mechanisms, similar to those which assist replication. RNA polymerase recognizes irregularities in the input chain (which register as structural deformations) and can reverse the synthesis process, cleave the incorrect nucleotide, or even terminate polymerization and discard the unfinished transcript. In the case of translation, the ribosome is generally incapable of verifying the correctness of the polypeptide chain due to encoding differences between the polymer and its template. The specificity of synthesis is associated with the creation of aa-tRNA and determined by the specific nature of the enzyme itself. Any errors introduced beyond this stage go uncorrected.

Once a polypeptide chain has been synthesized, it must fold in a prescribed way in order to fulfill its purpose. In theory, all proteins are capable of spontaneous folding and reaching their global energy minima. In practice, however, most proteins misfold and become "stuck" in local minima. (Alternatively, the global minimum may not represent the active form of the protein, and a local minimum may be preferable.) This is why the folding process itself is supervised by special regulatory structures: simple proteins (chaperones) or machine-like mechanisms (chaperonins), which work by attaching themselves to the polypeptide chain and guiding its folding process. Improperly folded structures are broken down via dedicated "garbage collectors" called proteasomes. If, however, the concentration of undesirable proteins reaches critical levels (usually through aggregation), the cell itself may undergo controlled suicide called apoptosis. Thus, the entire information pathway—starting with DNA and ending with active proteins—is protected against errors. We can conclude that fallibility is an inherent property of genetic information channels and that in order to perform their intended function, these channels require error correction mechanisms.

The processes associated with converting genetic information into biologically useful structures are highly complex. This is why polymerases (which are of key importance to gene expression) are large complexes with machine-like properties. Some of their subunits perform regulatory functions and counteract problems which may emerge in the course of processing genetic information. In the case of synthesis, the main goal of polymerases is to ensure equivalent synthesis of both anti-parallel DNA strands. In contrast, transcription relies on proper selection of the DNA fragment to be expressed. Problems associated with polypeptide chain synthesis usually arise as a result of difficulties in translating information from nucleic acids to proteins.

While the cell is in interphase its DNA is packed in the nucleus and assumes the form of chromatin. During cell division the nucleus is subdivided into so-called chromosome territories, each occupied by a pair of chromosomes. Densely packed fragments of DNA material, unavailable for transcription, constitute the so-called heterochromatin, while regions from which information can potentially be read are called euchromatin.

Expression of genetic information is conditioned by recognition of specific DNA sequences which encode proteins. The function of a protein is not, however, entirely determined by its coding fragment. Noncoding fragments may also store useful information, related to, e.g., regulatory mechanisms, structure of the chromatin strand (including its packaging), sites of specific aberrations such as palindromes, etc. Proper transfer of genetic information requires precise recognition of its nucleotide arrangement.

Specific DNA sequences may be recognized by RNA or by proteins. RNA recognition is straightforward, as both DNA and RNA share the same "language." Recognition of a nucleotide sequence by proteins poses more problems; however some proteins have evolved the ability to attach to specific sequences (transcription factors in particular). This mechanism is usually employed in the major groove of the DNA double helix whose breadth admits contact with transcription mediators (it should, however, be noted that the minor groove may also convey useful structural information via relative differences in its width and the distribution of electrostatic charges—see papers by T. Tullius and R. Rohs; 2009). All such information can be exploited by proteins in order to recognize their target sequences. Interaction between proteins and nucleotide sequences usually assumes the form of protein-RNA complexes which include a short double-stranded transcripts consisting of approximately 20–30 nucleotides. Such noncoding RNA fragments can bind as single-strand pieces to proteins which exhibit some defined properties (e.g., enzymes). Their presence ensures in the result that the protein acts in a highly targeted manner, seeking out sequences which correspond to the attached RNA strand. Examples include AGO RNase enzymes as well as proteins whose function is to inhibit or destroy outliving mRNA chains. They are thus important for maintaining biological balance within the cell. Protein-RNA complexes also participate in untangling chromatin strands, epigenetic processes, antiviral defense, and other tasks.

The double-stranded form assumed by RNA prior to interacting with proteins protects it from rapid degradation. Three groups of interfering RNA fragments have been distinguished with respect to their length, means of synthesis, and mechanism of action (they are called miRNA, siRNA, and piRNA, respectively). The use of short RNA fragments as "target guides" for proteins greatly increases the efficiency of functional expression of genetic information (Fig. 3.13). By the same token, specific RNA chains are also useful in translation processes. The ribosome is a nanomachine whose function depends on precise interaction with nucleic acids—those integrated in the ribosome itself as well as those temporarily attached to the complex during polypeptide chain synthesis (tRNA). Good cooperation between proteins and nucleic acids is a prerequisite of sustained function of the entire

**Fig. 3.13** Synthesis of complexes by using RNA fragments which guide active proteins to specific sites in the target sequence. The figure shows miRNA being used to degrade redundant mRNA

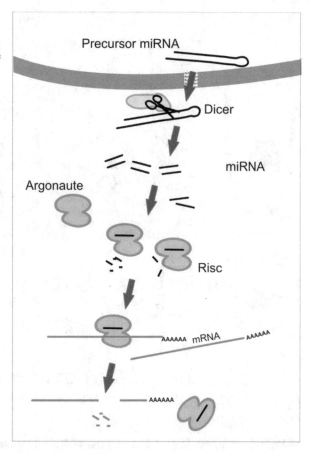

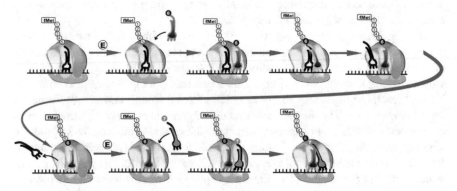

**Fig. 3.14** Translation process—the function of tRNA in recognizing mRNA and facilitating the synthesis of polypeptide chain

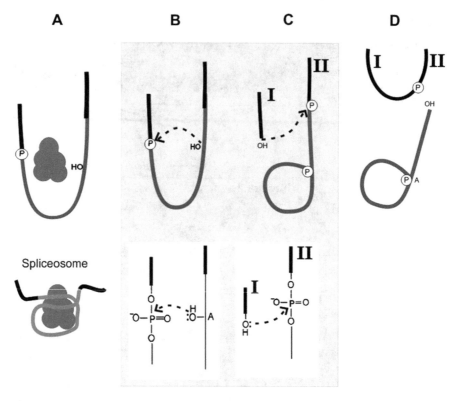

**Fig. 3.15**  I, splicing; II, alternative splicing

biological machinery (Fig. 3.14). The use of RNA in protein complexes is common across all domains of the living world as it bridges the gap between discrete and continuous storage of genetic information.

The discrete nature of genetic material is an important property which distinguishes prokaryotes from eukaryotes. It enables gene splicing in the course of transcription and promotes evolutionary development. The ability to select individual nucleotide fragments and construct sequences from predetermined "building blocks" results in high adaptability to environmental stimuli and is a fundamental aspect of evolution. The discontinuous nature of genes is evidenced by the presence of fragments which do not convey structural information (introns), as opposed to structure-encoding fragments (exons). The initial transcript (pre-mRNA) contains introns as well as exons. In order to provide a template for protein synthesis, it must undergo further processing (also known as splicing): introns must be cleaved and exon fragments attached to one another. The process is carried out by special complexes called spliceosomes, which consist of proteins and function-specific RNA fragments (Fig. 3.15). These fragments inform the spliceosome where to sever the pre-mRNA strand so that introns may be discarded and the remaining exon fragments reattached to yield the mRNA transcript template. Recognition of intron-exon boundaries is usually very precise, while the reattachment of adjacent exons is subject to some variability. Under certain conditions, alternative splicing may occur, where the ordering of the final product does not reflect the order in which exon sequences appear in the source chain. This greatly increases the number of potential mRNA combinations and thus the variety of resulting proteins. Alternative splicing explains the clear disparity between the number of genes in the genome and the variety of proteins encountered in living organisms. It also plays a significant role in the course of evolution. The discontinuous nature of genes is evolutionarily advantageous but comes at the expense of having to maintain a nucleus where such splicing processes can be safely conducted, in addition to efficient transport channels allowing transcripts to penetrate the nuclear membrane. While it is believed that at early stages of evolution RNA was the primary repository of genetic information, its present function can best be described as an information carrier. Since unguided proteins cannot ensure sufficient specificity of interaction with nucleic acids, protein-RNA complexes are used often in cases where specific fragments of genetic information need to be read.

Long RNA chains usually occur as single strands; however they can occasionally fold into double strands which resemble proteins and can even perform protein-like functions (including enzymatic catalysis, as observed in ribozymes). It should be noted, however, that catalytic activity has nothing in common with the natural RNA activity connected with sequence recognition, and hence both activities differs essentially.

In summary, we can state that the primary role of the genome and the information contained therein are to sustain living processes and enable cells to convey the mechanics of life to their offspring. This process depends on accurate expression of DNA information in the form of proteins and on their activity, maintaining the cell in a steady state what allows stabilizing biological processes in accordance with genetically programmed criteria.

### 3.2.3.1 Nuclear Pores

An example of a universal technical solution is provided by the so-called nuclear pores, which facilitate transport of substances between the nucleus and cell cytoplasm. This cross-membrane transport is highly diversified and must proceed without interruptions. Proteins are synthesized in the cytoplasm on the basis of RNA fragments synthesized and secreted by the nucleus. Therefore, mutual interaction between the nucleus and the cytoplasm is essential for the cell. Some of the most frequently synthesized proteins—ribosomal proteins—which emerge in the cytoplasm must return to the nucleus where they attach to RNA strands and then, as proper components of ribosomes, are transported back to the cytoplasm. Most molecules which migrate from the nucleus to the cytoplasm are mRNA transcripts; however, the nuclear membrane must also remain permeable for small metabolites. Taken together, these functions place special demands on transport channels. The corresponding transport mechanism must, on the one hand, remain universal, while on the other hand ensuring specificity and directionality (Fig. 3.16). The former is achieved by an elastic barrier composed of polypeptide fibrils which comprise

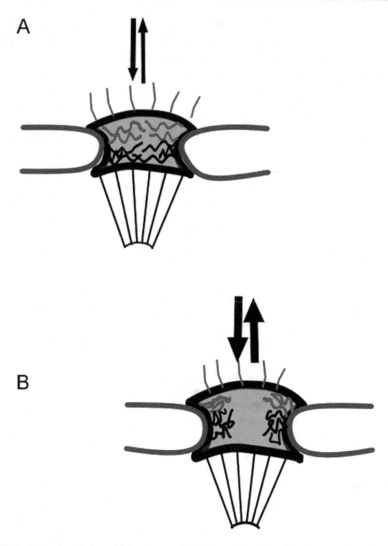

**Fig. 3.16** Systemic solution which ensures selectivity and diversity of nuclear membrane transport by employing fibrillar polypeptide structures called nucleoporins (FG/Nup)—a plastic barrier which can be modulated by appropriate signals depending on the needs and properties of transported molecules

mainly glycine and phenylalanine. Under the influence of the appropriate signals, these fibrils either expand or contract, creating gateways for substances which must be conveyed across the membrane. Transport is therefore controlled by specific signaling molecules and is an active process, requiring energy.

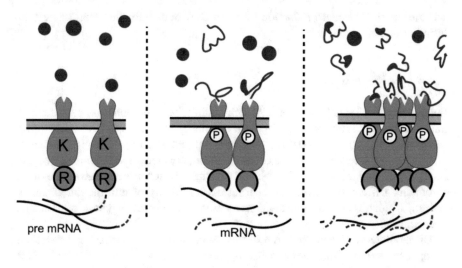

**Fig. 3.17**  Action of Ire1 receptors in the UPR system. Capture of misfolded proteins triggers an alarm signal. K, kinase; R, RNase

### 3.2.3.2 Protein Synthesis and Degradation

Maintaining protein homeostasis is of critical importance for cells, enabling them to function and respond to stressors. It is also important in the context of senescence and longevity of organisms.

Protein degradation is particularly a sensitive component of homeostasis because, unlike synthesis, it is inherently stochastic. Proteins differ with respect to longevity; what's more—misfolded or structurally deficient proteins may emerge, and the quantity of such proteins increases significantly under stress.

Proteins are degraded either by lysosomal proteases or via the proteasome pathway. Approximately 30% of all synthesized proteins are dispersed via the endoplasmic reticulum. Environmental conditions encountered in the reticulum are less controllable than in the cytoplasm—this results in locally greater levels of misfolded proteins whose conformational properties may promote unwanted aggregation. Consequently, the reticulum is equipped with a special receptor system which reacts to the presence and concentrations of undesirable proteins. This system is referred to as UPR (unfolded protein response). The role of UPR is to detect misfolded proteins and activate a stress response by which protein structure repair may occur (owing to chaperones) or, alternatively, the cell may undergo apoptosis. Some misfolded proteins may also penetrate into the cytoplasm, where they are destroyed in the proteasome pathway, through ubiquitination.

Overall, UPR consists of three components. The Ire1 receptor senses misfolded proteins (Fig. 3.17), and its action is linked to PERK kinase and the ATF6 transcription factor. UPR appears to play a key role in maintaining protein homeostasis. In the presence of misfolded proteins, Ire1 undergoes phosphorylation, while its constituent RNase becomes active. This, in turn, triggers an alarm through modification of a specific pre-mRNA and production of an mRNA matrix for the appropriate transcription factor.

### 3.2.4  Epigenetics

Epigenetics is a branch of science which studies the differentiation of hereditary traits (passed on to successive generations of cells by means of cell division) through persistent activation or inhibition of genes, without altering the DNA sequences themselves. Differentiation is a result of chemical, covalent modification of histones and/or DNA, and the action of non-histone proteins which affect the structure of the chromatin. Differentiation has no bearing on the fidelity of information channels; instead, it determines the information *content*, i.e., the set of genes released for transcription.

Differentiation plays a key role in the expression of specialized cell functions (as opposed to basic functions encoded by the so-called housekeeping genes, which are relatively similar in all types of cells). Information stored in DNA can be accessed by specific protein complexes which uncoil the chromatin thread and present its content for transcription. This process is guided by markers: modified (usually methylated) histone amino acids and/or methylated DNA nucleotides. Modifications ensure the specificity of binding between DNA and non-histone proteins and therefore guide the appropriate release of genetic information, facilitating biological development and vital functions. Intracellular differentiation processes are initiated at specific stages in cell development via RNA-assisted transcription factors. Their function can be controlled by external signals (hormones), capable of overriding intracellular regulatory mechanisms.

Epigenetic mechanisms are observed in the following processes:

1. Embryogenesis and regeneration
2. Stem cell survival and differentiation (e.g., bone marrow function)
3. Selective (single-allele) inheritance of parental traits (also called *paternal* and *maternal imprinting*) including functional inhibition of chromosome X
4. Epigenetics of acquired traits

Ad. 1. *Embryogenesis and Regeneration*

### 3.2.5  Development Genetics (Embryogenesis and Regeneration): The Principles of Cell Differentiation

Epigenetic differentiation mechanisms are particularly important in embryonic development. This process is controlled by a very specific program, itself a result of evolution. Unlike the function of mature organisms, embryonic programming refers to structures which do not yet exist but which need to be created through cell proliferation and differentiation. The primary goal of development is to implement

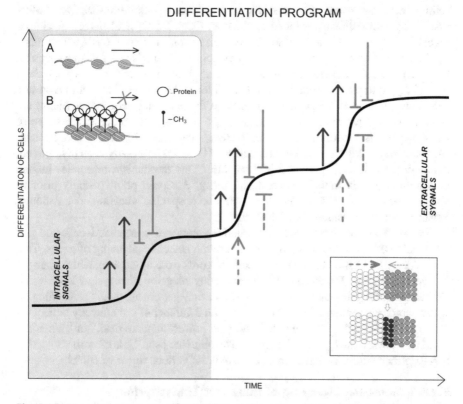

**Fig. 3.18** The sequential nature of cell differentiation. Arrows indicate intra- and intercellular signals which trigger each differentiation stage. Top inset: schematic depiction of a chromatin fragment (DNA + protein) in its active (**a**) and suppressed (**b**) form (structure and packing of complexes). Bottom inset: emergence of a differentiated group of cells (black layer) as a function of interaction between adjacent cell layers

the genetic blueprint—a task which is automated by sequential activation of successive development stages in accordance with chemical signals generated at each stage (Fig. 3.18). Similar sequential processing can be observed in cell division which consists of multiple, clearly defined stages. It should be noted that embryonic development programs control both *proliferation* and *differentiation* of cells.

Differentiation of cells results in phenotypic changes. This phenomenon is the primary difference between development genetics and steady-state genetics. Functional differences are not, however, associated with genomic changes: instead they are mediated by the *transcriptome* where certain genes are preferentially selected for transcription, while others are suppressed.

In a mature, specialized cell, only a small portion of the transcribable genome is actually expressed. The remainder of the cell's genetic material is said to be *silenced*. Gene silencing is a permanent condition. Under normal circumstances mature cells never alter their function, although such changes may be forced in a laboratory setting, e.g., by using viral carriers to introduce special transcription factors associated with cellular pluripotency (Nanog, Oct4, Sox2, Klf4, Myc). A similar reversal of aging processes is also observed in neoplastic tissue. Cells which make up the embryo at a very early stage of development are *pluripotent*, meaning that their purpose can be freely determined and that all of their genetic information can potentially be expressed (under certain conditions). Maintaining the chromatin in a change-ready state is a function of hormonal factors called morphogens (e.g., the *sonic hedgehog* protein). At each stage of the development process, the scope of pluripotency is reduced until, ultimately, the cell becomes monopotent. Monopotency implies that the final function of the cell has already been determined, although the cell itself may still be immature. This mechanism resembles human education which is initially generalized, but at a certain point (usually prior to college enrollment), the student must choose a specific vocation, even though he/she is not yet considered a professional.

As noted above, functional dissimilarities between specialized cells are not associated with genetic mutations but rather with selective silencing of genes. This process may be likened to deletion of certain words from a sentence, which changes its overall meaning. Let us consider the following adage by Mieczysław Kozłowski: "When the blind gain power, they believe those they govern are deaf." Depending on which words we remove, we may come up with a number of semantically nonequivalent sentences, such as "When the blind gain power, they are deaf." or "When the blind gain power, they govern." or even "The blind are deaf." It is clear that selective transcription of information enables us to express various forms of content.

### 3.2.5.1 Molecular Licensing of Genes for Transcription

The "gene licensing" mechanism depends primarily on chemical modifications (mostly methylation of histones but also of DNA itself) and attaching the modified chromatin to certain non-histone proteins. In addition to methylation, the activity of a given gene may be determined by acetylation, ubiquitination, sumoylation, and phosphorylation. However, the nature of the modifying factor is just one piece of the overall puzzle. Equally important is the modification site: for instance, methylation of lysine at position 4 of histone 3 (H3K4) promotes transcription, while methylation of lysine at position 9 (H3K9) results in a different DNA-protein binding and inhibits gene expression (Fig. 3.19). The degree of methylation (the presence of one, two, or three methyl groups) matters as well: triple methylation usually occurs at positions 4, 9, 27, and 36 of histone 3, as well as at position 20 of histone 4. Positions 9 and 27 are particularly important for gene suppression because methylated lysine acts as an acceptor for certain *Polycomb* proteins which inhibit transcription. On the other hand, position 4 of histone 3 is associated with promotion of gene transcription mediated by *trithorax* proteins.

Gene suppression may be reversed through detachment and/or demethylation of the coupled proteins (Fig. 3.20a). Phosphorylation associated with introduction of a negative charge promotes dissociation of inhibitors and is therefore useful in regulatory mechanisms (Fig. 3.20b). It is interesting to note that phosphorylation may also occur in histones: for instance, histone 3 includes serine units which bind

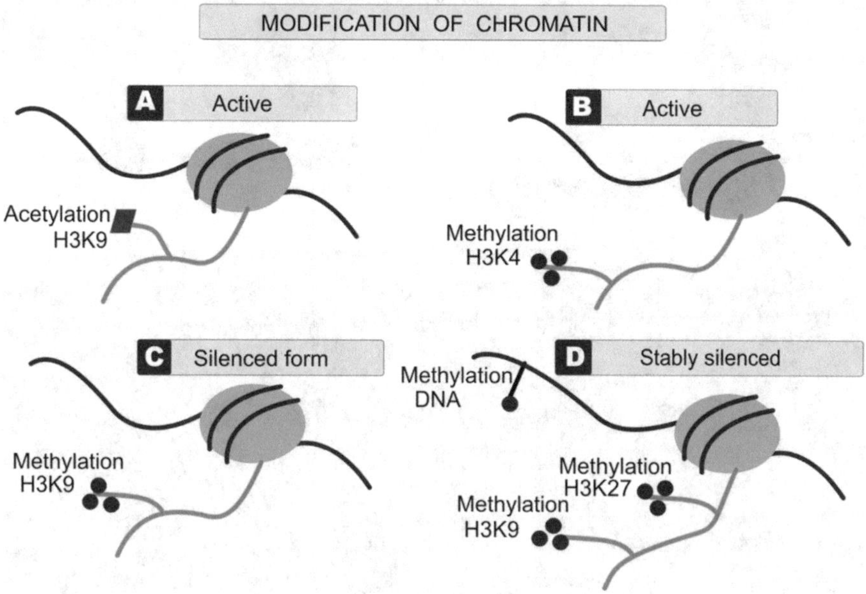

MODIFICATION OF CHROMATIN

**Fig. 3.19** Simplified model of histone and DNA methylation and its results

phosphoric acid residues at positions 10 and 28, i.e., directly adjacent to the lysine units at positions 9 and 27, which (as noted above) inhibit transcription by binding with additional proteins. Serine phosphorylation in histones controls the bonding between DNA and non-histone proteins (which may also undergo phosphorylation). On the other hand, removal of methyl groups is a function of specific demethylases.

In addition to lysine, histone methylation may also affect arginine, while direct DNA methylation usually involves cytosine. Such modifications are often mutually dependent (Fig. 3.21). Methylation is a rapid, covalent process and a convenient way to directly tag the DNA chain as the replication complex progresses. It enables cells to pass epigenetic information to their offspring and thus ensures its persistence (Fig. 3.22). However, the final decision on which genes to silence is a function of non-histone proteins which bind to DNA in places marked via methylation (or other forms of chemical modification).

In contrast to methylation, acetylation usually promotes transcription. Many transcription activators are in fact enzymes (acetyltransferases), and, consequently, many gene suppressor proteins act by deacetylation. Non-histone proteins involved in epigenetic processes either activate or inhibit transcription and may also modify the structure of chromatin. DNA–protein complexes which act as gene suppressors (for instance, those involving *Polycomb* proteins) result in tighter packing of the heterochromatin chain (the silent, non-transcribed part of the genome). Heterochromatin may be packed in at least four different ways, depending on the activity of attached proteins; however we usually distinguish two broad categories: constitutive (permanently suppressed) heterochromatin and facultative heterochromatin, which

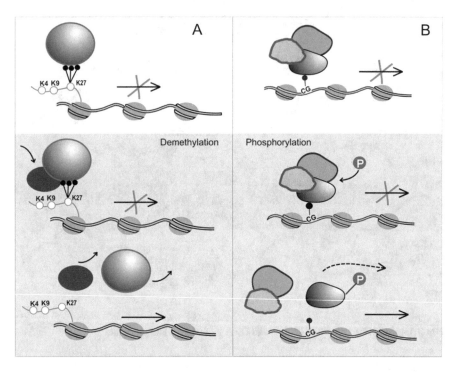

**Fig. 3.20** Chromatin activation (dissociation of non-histone proteins) through (**a**) demethylation of histone H3K27me3 and (**b**) protein phosphorylation (schematic depiction)

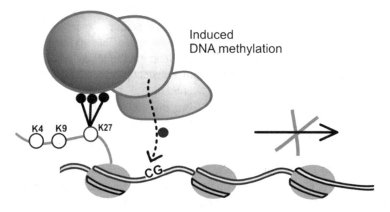

**Fig. 3.21** Schematic depiction of the interactions between a methylated histone (H3K27me3) and DNA

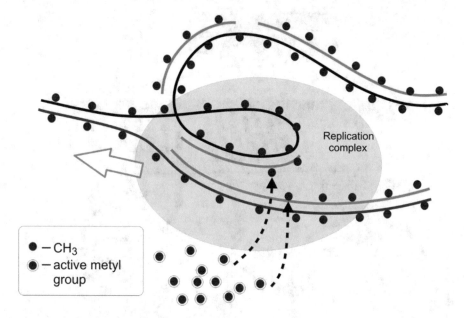

**Fig. 3.22** Simplified view of remethylation of freshly synthesized genetic material (DNA and histones) by the replicating complex and its methylating subunits

may, under certain condition, be expressed (note that the structure which allows transcription of genetic material is called euchromatin).

### 3.2.5.2 Specificity of Epigenetic Processes

The complicated chromatin modification mechanism associated with cellular differentiation is a consequence of the complex selection of genes which need to be expressed or silenced in order to sustain biological activities at the proper level. It can be compared to a piano concerto where the pianist must strike certain keys in a selected order at just the right time. Moreover, each stroke must have an appropriate force, determining the volume and duration of the played note. Some keys are struck separately, while others need to be arranged into chords. All of these decisions are subject to a form of programming, i.e., the notes written down by the composer of a given piece.

The specificity of recognition and activation of certain nucleotide sequences for transcription seems understandable if we assume that genes may be recognized by transcription factors alone or in collaboration with noncoding RNA fragments which belong to the miRNA group. Both types of structures are capable of interfacing with DNA regulatory sequences and thus selectively induce transcription of certain genes. However, contrary to a piano concerto where the role of the pianist is simply to play back the piece by striking certain keys only, the cell must also proactively manage its silenced genes.

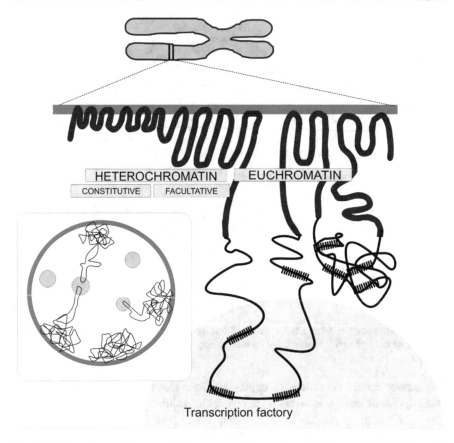

HETEROCHROMATIN          EUCHROMATIN

CONSTITUTIVE     FACULTATIVE

Transcription factory

**Fig. 3.23** Spatial restrictions applied to transcription and methylation of DNA (chromatin fragments) through selection of areas where enzymatic contact is maintained. Inset: depiction of the nucleus

Chromatin methylation and other chemical modifications are a result of enzymatic activity where the substrates (basal histone amino acids and cytosine of DNA) reside both within the transcribed parts of DNA and in sections which need to be silenced. Clearly, this property may interfere with the selectivity of gene expression.

A solution emerges in the form of spatial isolation of certain DNA fragments and exposes selected parts of the chain for enzymatic activity. This is only possible during interphase when transcription and other enzymatic processes appear to be concentrated in specific areas of the nucleus (sometimes called *factories*). These areas accept "loosened" DNA coils, recognized and preselected by transcription factors and/or RNA. Compartmentalization also prevents uncontrolled propagation of catalysis (Fig. 3.23).

Spatial ordering of catalysis is important for epigenetic processes due to the great variety of enzymatic interactions involved in cell differentiation. However, an even more important self-control mechanism associated with enzymatic activity seems to be its division into stages, where only selected types of enzymes seem to be active at each stage. This greatly increases the selectivity of information channels and reduces the potential for error.

### 3.2.5.3 External Control of Cell Proliferation and Differentiation: Embryonic Development

Each stage of differentiation can be activated automatically; however they all obey steering signals which come from outside of the cell, i.e., from other cells. Such signals can be generated directly (by adjacent cells) or indirectly (by specific hormonal markers called morphogens).

The duration of the signal and the concentration of a specific morphogen may affect cell differentiation by triggering internal processes which subsequently operate in accordance with predetermined sequential programs. Morphogen diffusion is, however, somewhat peculiar: morphogens travel through clusters of densely packed embryonic cells and have to maintain a predetermined concentration at a given distance from their originator in order to ensure proper strength of the signal they encode. In order to fulfill these goals, morphogens are inherently short-lived and need to be constantly replenished. They must also possess special means of traversing cell clusters. It should be noted that the boundaries separating various tissues are usually well delineated in spite of the diffusive nature of biological signals. This is due to simultaneous action of contradictory signals, which results in the emergence of unambiguous tissue boundaries (Fig. 3.24).

Each gene packet activated in the course of differentiation belongs to a certain development stage. This alignment results in staged synthesis of various sets of proteins and enzymes, each responsible for performing different actions. For

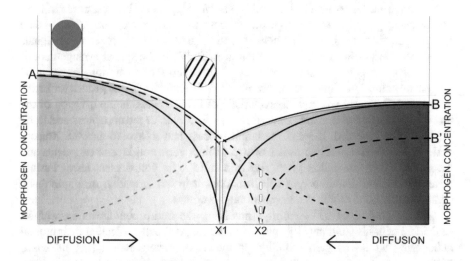

**Fig. 3.24** Formation of a clear boundary between separate cell layers in an evolving embryo, through contradictory action of morphogens. Circles represent various types of specialized cells associated with different concentrations of a given morphogen

instance, in stem cells methylation affects CG and CA nucleotide clusters, whereas in mature, specialized cells, no CA methylation is observed. Furthermore, as no cell is completely independent of its adjacent cells, cellular development must proceed in a coordinated fashion.

Coordinated propagation of information follows a hierarchical pathway, meaning that information first reaches key loci in the developing system and only then can be disseminated to wider groups of recipients (genes). This process resembles a human population settling a new territory: initially, settlers decide upon administrative boundaries and elect local authorities. Later on, these agreements may be amended as a result of individual postulates and specific strategies developed in order to resolve emerging problems and adapt to changing conditions.

Cells responsible for triggering new signal pathways need to be created in the course of embryonic development. The emergence of new centers of activity and new tissues results from cooperation of existing, differentiated cells. Cell proliferation and mutual interactions proceed in accordance with the genetic program, progressively giving rise to new structures. The spatial "blueprint" of the embryonic mass is in place even before macroscopic details can be discerned; indeed, the process of differentiation begins with the first asymmetric division of the embryo. Cells which are already undergoing differentiation "remember" their position and place in the development program—thus, they can be said to possess a specific "address" in the overall structure of the organism. The placement of cells in a developing embryo is determined by the *hox* gene family. In humans, these genes (of which there are approximately 40) are activated sequentially during successive development stages. Their action is to impose spatial alignment upon the growing mass of cells. Spatial memory and information about the cell's future role in the developing organism are stored in its chromatin, conditioned to enable certain types of transcription. Generalized biochemical signals trigger specific responses in individual cells, guiding the development process in each part of the embryo. Cell groups gain their epigenetic "addresses" and "assignments" by reacting to signals in different ways, thus enabling coherent growth. While the spatial arrangement of tissues was subjected to some changes in the course of evolution, the general epigenetic control mechanisms governing the differentiation of cell groups remained unchanged. An early strategy, characteristic of invertebrates, is the division of the embryonic mass into segments (see Fig. 3.34). However, as the notochord and (later on) the spine emerged, the differentiation process had to evolve as well. Thus, a vertebrate embryo initially consists of two distinct germ cell layers: endoderm and ectoderm. Their interaction gives rise to a third layer called mesoderm. Further development and divergence of cell layers result in the formation of a spine as a central core around which development may progress.

Genetic control of cell mobility (involving the entire cell layers as well as individual cells) is facilitated by changes in the shape of cells, affecting their mutual adhesion. Cell layers gain mobility by means of locally reduced or increased adhesion, itself a result of the emergence or degradation of surface receptors (cadherins and integrins).

Other examples of epigenetic mechanisms are as follows:

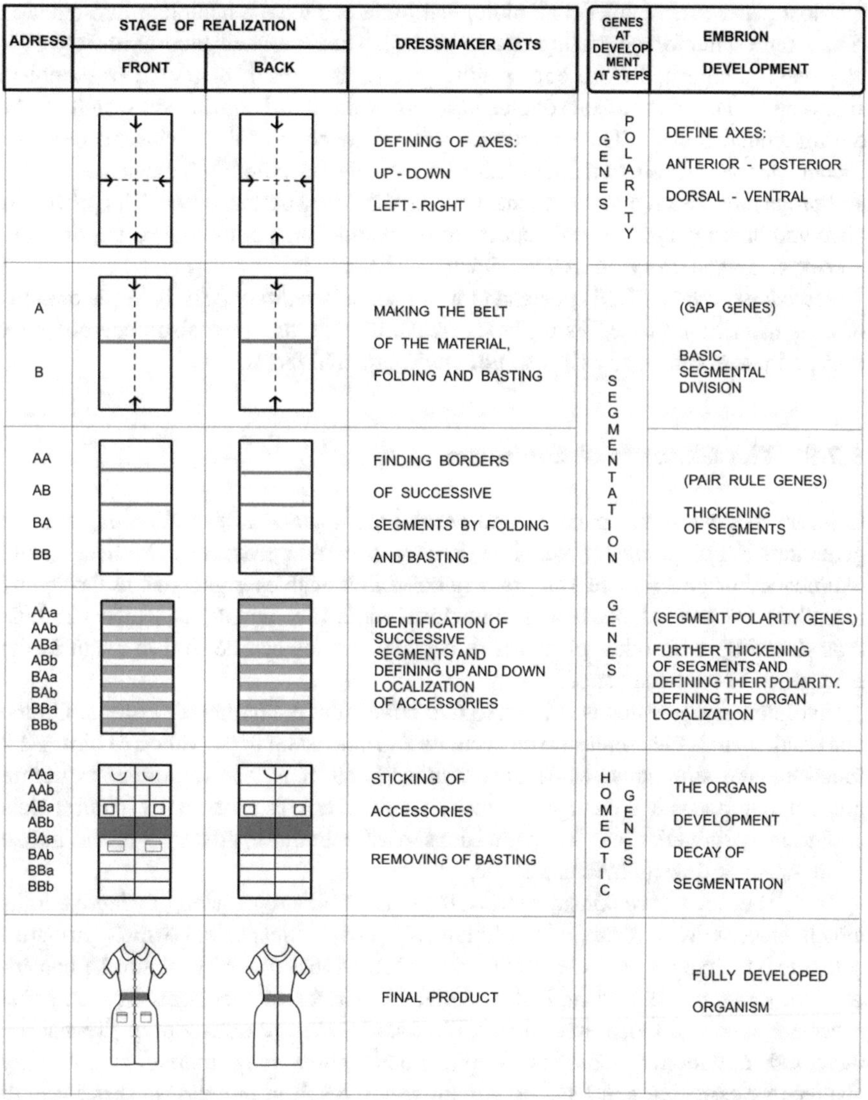

| ADRESS | STAGE OF REALIZATION | | DRESSMAKER ACTS | GENES AT DEVELOPMENT AT STEPS | EMBRION DEVELOPMENT |
|---|---|---|---|---|---|
| | FRONT | BACK | | | |
| | | | DEFINING OF AXES: UP - DOWN LEFT - RIGHT | GENES POLARITY | DEFINE AXES: ANTERIOR - POSTERIOR DORSAL - VENTRAL |
| A B | | | MAKING THE BELT OF THE MATERIAL, FOLDING AND BASTING | S E G M E N T A T I O N | (GAP GENES) BASIC SEGMENTAL DIVISION |
| AA AB BA BB | | | FINDING BORDERS OF SUCCESSIVE SEGMENTS BY FOLDING AND BASTING | | (PAIR RULE GENES) THICKENING OF SEGMENTS |
| AAa AAb ABa ABb BAa BAb BBa BBb | | | IDENTIFICATION OF SUCCESSIVE SEGMENTS AND DEFINING UP AND DOWN LOCALIZATION OF ACCESSORIES | GENES | (SEGMENT POLARITY GENES) FURTHER THICKENING OF SEGMENTS AND DEFINING THEIR POLARITY. DEFINING THE ORGAN LOCALIZATION |
| AAa AAb ABa ABb BAa BAb BBa BBb | | | STICKING OF ACCESSORIES REMOVING OF BASTING | HOMEOTIC GENES | THE ORGANS DEVELOPMENT DECAY OF SEGMENTATION |
| | | | FINAL PRODUCT | | FULLY DEVELOPED ORGANISM |

**Fig. 3.34** Points of reference in a developing embryo compared to the work of a blind tailor

### Ad. 2. *Stem Cell Survival and Differentiation*

A special group of undifferentiated cells, called stem cells, may persist in mature organisms in specific niches formed by adjacent cell layers. One example of such a structure is bone marrow, where new blood cells are constantly being created.

### Ad. 3. *Selective Inheritance of Parental Trials*

*Imprinting*—Differentiation mechanisms can also be used to ensure monoallelicity.

Most genes which determine biological functions have a biallelic representation (i.e., a representation consisting of two alleles). The remainder (approximately 10% of genes) is inherited from one specific parent, as a result of partial or complete silencing of their sister alleles (called *paternal* or *maternal imprinting*) which occurs during gametogenesis. The suppression of a single copy of the X chromosome is a special case of this phenomenon. It is initiated by a specific RNA sequence (XIST) and propagates itself, eventually inactivating the entire chromosome. The process is observed in many species and appears to be of fundamental biological importance.

Ad. 4. *Epigenetics at Acquired Trials*

*Hereditary traits*—Cell specialization is itself a hereditary trait. New generations of cells inherit the properties of their parents, though they may also undergo slight (but permanent) changes as a result of environmental factors.

## 3.2.6  The Genetics of Evolution

Contrary to steady-state genetics and development genetics, evolution exploits the gene mutation phenomenon which underpins speciation processes. Duplication and redundancy of genetic material are beneficial as it enables organisms to thrive and reproduce in spite of occasional mutations. Note that mutations which preclude cross-breeding with other members of a given species can be said to result in the emergence of a new species.

Evolutionary genetics is subject to two somewhat contradictory criteria. On the one hand, there is clear pressure on accurate and consistent preservation of biological functions and structures, while on the other hand, it is also important to permit gradual but persistent changes. Mutational diversity is random by nature; thus evolutionary genetics can be viewed as a directionless process—much unlike steady-state or developmental genetics.

In spite of the above considerations, the observable progression of adaptive traits which emerge as a result of evolution suggests a mechanism which promotes constructive changes over destructive ones. Mutational diversity cannot be considered truly random if it is limited to certain structures or functions. In fact, some processes (such as those associated with intensified gene transcription) reveal the increased mutational activities. Rapid transcription may induce evolutionary changes by exposing cell DNA to stimuli which result in mutations. These stimuli include DNA repair processes, particularly those which deal with double-strand damage. In this respect, an important category of processes involves recombination and shifting of mobile DNA segments.

Approximately 50% of the human genome consists of mobile segments, capable of migrating to various positions in the genome. These segments are called transposons and retrotransposons (respectively, DNA fragments and mobile RNA transcripts which resemble retroviruses in their mechanism of action except that they are not allowed to leave the cell).

The mobility of genome fragments not only promotes mutations (by increasing the variability of DNA) but also affects the stability and packing of chromatin strands wherever such mobile sections are reintegrated with the genome. Under normal circumstances the activity of mobile sections is tempered by epigenetic mechanisms (methylation and the DNA-protein complexes it creates); however in certain situations, gene mobility may be upregulated. In particular, it seems that in prehistoric times such events occurred at a much faster pace, accelerating the rate of genetic changes and promoting rapid evolution.

Cells can actively promote mutations by way of the so-called AID process (*activity-dependent cytosine deamination*). It is an enzymatic mechanism which converts cytosine into uracil, thereby triggering repair mechanisms and increasing the likelihood of mutations. AID is mostly responsible for inducing hypermutations in antibody synthesis, but its activity is not limited to that part of the genome. The existence of AID proves that cells themselves may trigger evolutionary changes and that the role of mutations in the emergence of new biological structures is not strictly passive.

### 3.2.6.1 Combinatorial Changes as a Diversity-Promoting Strategy

Although the processes mentioned above may contribute to evolutionary changes and even impart them with a certain direction, they remain highly random and thus unreliable. A simple increase in the rate of mutations does not account for the high evolutionary complexity of eukaryotic organisms. We should therefore seek an evolutionary strategy which promotes the variability of DNA while limiting the randomness associated with mutations and preventing undesirable changes.

This problem may be highlighted by considering the immune system which itself must undergo rapid evolution in order to synthesize new types of recombinant proteins called antibodies. As expected, antibody differentiation is subject to the same deterministic mechanisms which have guided evolution throughout its billion-year course but which remain difficult to distinguish from stochastic evolutionary processes. In the immune system, synthesis of new proteins (i.e., antibodies with new V domains) proceeds by way of changes in amino acid sequences (particularly in their V, D, and J DNA segments) through a mechanism which owes its function to high redundancy of certain fragments of genetic code. DNA sequences which contain the previously mentioned V, D, and J segments may, upon recombination, determine the structure of variable immunoglobulins: their light (V, J) and heavy chains (V, D, J) (see Fig. 3.36).

A key advantage of recombination is that it yields a great variety of antibodies, making it likely that an antibody specific to a particular antigen will ultimately be synthesized. The degree of variability in L and H chains is determined by the number of possible combinations of V/J and V/D/J segments (for L and H chains, respectively). Constructing random genetic sequences via recombination is a process which may occur far more frequently than creating new, complete genes from scratch. It enables great genetic diversity in spite of the limited participating genome information and is therefore preferable to gene differentiation. Antibody differentiation also relies on one additional mechanism which triggers random changes in their active groups: combination of light and heavy chains within the antigen binding site.

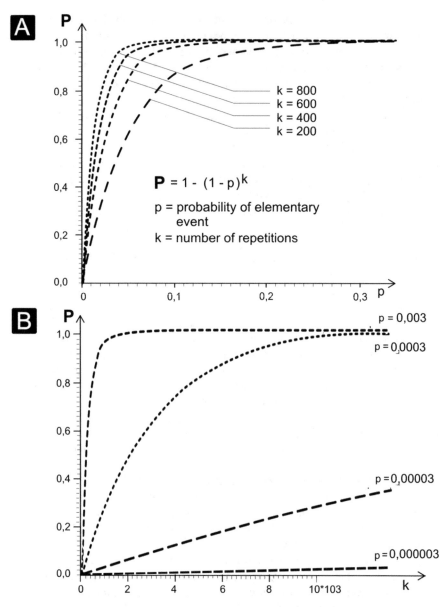

**Fig. 3.36** Association between the probability of hitting the target (*P*) and (**a**) the probability associated with each elementary event ( *p*) for a variable number of attempts (*k*) and (**b**) the number of attempts (*k*) for a variable elementary probability ( *p*)

Combinatorial differentiation and antibody synthesis may roughly be compared to the work of a cook who has to prepare meals for a large group of gourmands. Two strategies may be applied here: (1) preparing a large number of varied meals, far

**Table 3.2** Variability of antibodies as a result of recombination in V, D, and J segments and interaction between light (L) and heavy (H) chains

|   | SEGMENT | H | L κ | L λ |
|---|---|---|---|---|
| A | V | 250–1000 | 125–250 | 1–2 |
|   | J | 1–4 | 1–4 | 1–3 |
|   | D | 3–12 | 0 | 0 |
|   | SEGMENT | H | L κ | L λ |
|   | V | 250 | 125 | 1 |
|   | J | 1 | 1 | 1 |
|   | D | 3 | 0 | 0 |
| B | V x J x D | 3000 | 250 | 2 |
|   | H x Lκ | 750 000 |   |   |
|   | H x Lλ | 6000 |   |   |
|   | TOTAL | $4{,}5 \times 10^9$ |   | 387 |
|   | SEGMENT | H | L κ | L λ |
|   | V | 1000 | 250 | 2 |
|   | J | 4 | 4 | 3 |
|   | D | 12 | 0 | 0 |
| C | V x J x D | 48 000 | 1000 | 6 |
|   | H x Lκ | 48 000 000 |   |   |
|   | H x Lλ | 288 000 |   |   |
|   | TOTAL | $1{,}382 \times 10^{13}$ |   | 1 275 |

A, number of elements participating in recombination; B and C, values corresponding to smallest (B) and largest (C) sets of participating elements. Grayed-out fields represent cases where recombination is not involved

more than there are customers, and (2) preparing a selection of meal components (A, main courses; B, salads; C, appetizers; etc.) and allowing customers to compose their own sets. Clearly, the latter solution is more efficient and corresponds to strategies which can frequently be observed in nature. The efficiency of combinatorial differentiation is shown in Table 3.2, which presents a quantitative example of constructing antibodies from segments of the heavy chain (H) and two forms of the light chain (Lλ and Lκ). The degree of variability of each form is listed in Table 3.2A. As can be observed, this variability is far greater than in the case of a single, nonrecombinant chain consisting of all the above-mentioned segments.

The number of possible antibody sequences, given minimal variability of individual components, is listed in Table 3.2B. Table 3.2C presents values which correspond to the highest possible variability of components.

Combinatorial differentiation yields a huge population of immunoglobulins, making it exceedingly likely that at least some of them will be selectively adapted to their intended purpose. Further structural improvements are possible as a result of hypermutations restricted to the active protein group and induced by the AID process, resulting in incremental synthesis of more specialized antibodies (affinity maturation). Thus, antibody synthesis is itself a microscale model of directed evolution, enabling progressive improvement of its final product.

Thus although mutations are the basic mechanism by which changes in nucleotide sequences (and, consequently, amino acid sequences) can be introduced, variability of antibodies is not entirely dependent on mutations. Rather, it is a result of recombinant synthesis of diverse DNA fragments, each contributing to the structure of the final product. Such recombinant fragments emerge in the course of evolution, mostly via duplication of genetic code and also (in a limited scope) via localized mutations which do not affect the fundamental properties of the protein complex. The entire process is similar to a tool with exchangeable parts, although in the case of antibodies there is no "set of available parts"—rather, the given parts are synthesized on the fly from smaller subcomponents.

Combinatorial rearrangement of presynthesized DNA fragments (as opposed to ad hoc mutations) is an evolutionarily favored means of achieving diversity. It can rapidly accelerate DNA diversification while restricting the likelihood of adverse changes and errors associated with random mutations. The mechanism can be compared to the use of numbers and letters in car license plates, which also affords a great number of unique combinations.

In DNA, combinatorial diversification requires that functional fragments of the chain be clearly separated and well-spaced. Long dividers carry information which enables proper folding of the chain and assists in its combinatorial rearrangement. The discontinuing of genetic code supports combinatorial genetics but also facilitates the current gene expression through alternative splicing of exon fragments whenever suitable mRNA chains need to be synthesized. Thus, the number of intracellular proteins far exceeds the number of individual genes which make up the genome. This phenomenon is similar to recombination, although it applies to RNA rather than to DNA. Alternative splicing, itself a result of evolution, is an important contributor to evolutionary progress. Discontinuity is also observed in the so-called cis-regulatory elements of the genetic code which may be located far away from gene promoters. Such fragments include *enhancers* and *silencers*, separated by special sections called *insulators*.

DNA fragments, recognized by transcription factors, can bind to polymerase and guide its activity (Fig. 3.25). Regulatory fragments act as hooks for transcription factors. Sets of genes associated with a single biological function often share identical (or similar) enhancers and silencers, acted upon by a single transcription factor. Such cooperation of genes can be compared to piano chords which consist of several different notes but are struck by a single hand. The role of regulatory sequences in evolutionary development is more significant than that of actual protein codons (exons).

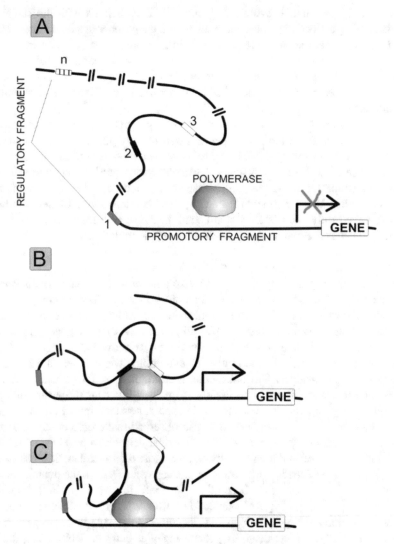

**Fig. 3.25** Differentiation complex synthesis (action of transcriptase). 1, 2, 3, …, $n$, DNA enhancers and silencers. Transcription factors not shown

Primitive organisms often possess nearly as many genes as humans, despite the essential differences between both groups. Interspecies diversity is primarily due to the properties of regulatory sequences. Evolutionary development promotes clear separation of DNA fragments carrying information concerning structure and function, allowing genetic code to be recombined with ease. In humans the separators between coding and noncoding DNA sequences (introns and exons) are among the longest observed in any organism. It therefore appears likely that diversification of regulatory structures carries significant evolutionary benefits.

As already mentioned, evolutionary progress is associated with the scope and diversity of regulatory sequences rather than with the number of actual genes. This is due to the fact that regulatory sequences facilitate optimization of gene expression. Returning to our metaphor, we can say that the same grand piano can be used either by a master pianist or by an amateur musician, although in the latter case the instrument's potential will not be fully realized, resulting in a lackluster performance.

The special evolutionary role of regulatory fragments is a consequence of their noncoding properties. Contrary to genes, noncoding fragments are not subject to structural restrictions: they do not need to be verified by the synthesis of specialized proteins where mutations are usually detrimental and result in negative selection. They also exhibit far greater variability than gene-encoding fragments. The mutability of noncoding DNA fragments is aided by the fact that—owing to their number— each fragment only contributes a small share to the overall regulatory effect. This property reduces the potential impact of unprofitable mutations.

### 3.2.6.2 Directed Mutability: Hotspot Genes

The recombinant variability of regulatory fragments and of genes themselves is sufficient to explain the progress of evolution. Nevertheless, ongoing research suggests the existence of additional mechanisms which promote evolution by increasing mutability in focused and localized scopes. Not all genes are equally susceptible to evolutionary pressure. Some can be termed "conservative" (i.e., undergoing few changes in the course of evolution), while others are subject to particularly rapid changes. The latter group is colloquially said to consist of *hotspot genes*. The reason behind this variability is unclear; however it appears that high mutability may emerge as a result of intense functional involvement or local instabilities in chromatin structure. It is also observed that fragments directly adjacent to retrotransposons are characterized by relatively loose packing, which may accelerate the rate of mutations. However, the most likely explanation has to deal with the presence of special nucleotide sequences which reduce the overall stability of the DNA chain. Certain observations attribute this role to short fragments dominated by a single type of nucleotide (usually T or A) attached to longer sequences which are largely bereft of nucleosomes. Such structures are particularly conductive to random exchange of genetic material between DNA coils, thereby promoting recombination and increasing the rate of mutations.

If such accelerated mutability is restricted to specific DNA fragments, its destructive impact can be minimized, and the mechanism may serve a useful purpose. It seems that the placement of such sequences in the genome may assist in directed evolutionary development. This is somewhat equivalent to the hypermutation process in antibody synthesis, where increased mutational activity (caused by AID) only applies to specific domains, ensuring an effective immune response without significantly altering the core structure of antibodies.

### 3.2.6.3 Gene Collaboration and Hierarchy

The placement of a given gene in the gene regulatory network may affect its transcriptional activity. Each mechanism which contributes to the overall phenotype requires the collaboration of many genes. The goal of such collaboration is to ensure balanced responses to various stimuli and activate all the required genes. Collaborative systems emerge via coupling of genes which together determine biological signals associated with transcription. Coherent activation of genes coding factors is a prerequisite for the formation of a so-called kernel.

Automatically regulated collaborative systems may be likened to cybernetic mechanisms. The impact of individual genes on collaboration processes is, however, unequal. Genes which occupy core nodes of regulatory networks (also called *input/output genes*) are usually tasked with proper routing of biological signals. Their intense functional involvement and interaction with advanced regulatory mechanisms may result in increased susceptibility to mutations.

Regulatory mechanisms which assist in evolutionary development are themselves subject to evolution—for instance, through creation of new enhancers and silencers or by increasing their relative spacing (similarly to introns). Gene regulation and interaction (particularly in the scope of input/output genes) can also be improved. Finally, the number of genes which encode transcription factors tends to increase over time. Such changes can be explained by their positive effect on gene collaboration. Referring to our "musical" example, we can say that using all ten fingers gives the pianist far greater leeway than if he were to tap the melody with just one finger.

As mentioned above, the role of genes in collaborative systems differs from gene to gene. Input/output genes are particularly important: it seems that they are the key members of the so-called *hotspot gene* set. This observation is further supported by their high involvement in transcription processes. It is theorized that the placement of input/output genes in the genome is intimately tied to their evolutionary role. An unambiguous proof of this theory would further confirm the directed nature of evolution. Mechanisms which accelerate evolutionary development (such as duplication, recombination based on discontinuity of genetic material and focused mutability) indicate that evolutionary processes follow specific strategies which may themselves undergo improvement. This, in turn, suggests the selective possibility of accelerating evolution. One putative example of this phenomenon is the rapid development of the *Homo sapiens* brain, often described as an "evolutionary leap."

## 3.3 Types of Information Conveyed by DNA

Biochemistry explains how genetic information can be used to synthesize polypeptide chains. On the surface it might appear that information transferred to RNA and subsequently to proteins is the only type of information present in the genome. However, an important issue immediately emerges: in addition to *what* is being synthesized, the living cell must also be able to determine *how*, *where*, *when*, and even *to what extent* certain phenotypic properties should be expressed.

**Fig. 3.26** Schematic depiction of a negative feedback loop, with elements determining the *how much?* and *how?* properties. The *what?* property may relate to each structure separately or to the loop as a whole

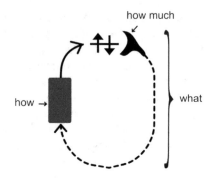

The questions *what?* and *when?* involve structural properties, while terms such as *how?* and *how much?* are more closely tied to function. The question *how?* usually emerges whenever we wish to determine the role of a certain structure, its synthesis process, or its mechanism of action. Each of these aspects may also be associated with the question *how much?*, i.e., a request for quantitative information. This information is useful in determining the required concentrations of reagents, their level of activity, the size of biological structures, etc. Quantitative assessment is important for any doctor who sends a biological sample to a diagnostics lab. Such properties are static and must therefore have a genetic representation. If we assume that, on a molecular level, structure determines function, we must also accept that the structure of certain proteins determines their quantity.

As can be expected, quantitative regulation is a function of receptor structures, each of which belongs to a regulatory chain. The question *how much?* is inexorably tied to regulatory processes. It seems clear that the stable concentrations and levels of activity observed in biological systems cannot be maintained without regulation. It is equally evident that such regulation must be automatic since isolated cell cultures can thrive and maintain their biological properties despite not being part of any organism.

Research indicates that biological regulatory mechanisms rely basically on negative feedback loops. Figure 3.26 presents the structure of such a loop. The heredity of a biological function is not restricted to the specific effector which directly implements it but covers the entire regulatory chain, including receptor systems and information channels.

Biological function cannot be separated as long as it is in the range of physiological regulation.

Quantitative control of the activity of various processes is facilitated by receptors which measure product concentrations or reaction intensity. Each receptor is connected to an effector which counteracts the observed anomalies. The receptor—being a functional protein—contains an active site which it uses to form reversible complexes with elements of the reaction it controls (usually with its products).

Typical receptors are allosteric proteins which undergo structural reconfiguration and release a signal whenever a ligand is bound in their active site. The equilibrium constant—a measure of receptor-ligand affinity—determines the concentration at

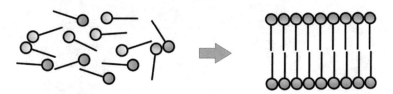

**Fig. 3.27**  Self-organization example: membrane formation

which saturation occurs and the receptor morphs into its complementary allosteric form. As a consequence, ligand concentration depends on the affinity of its receptor and therefore on its structure. The structure of the receptor protein determines the quantitative properties of the system as a whole, thus providing an answer to the *how much?* question. In contrast, the *how?* issue is addressed by the structure of effector proteins.

Effectors may be either simple or complex, depending on the task they perform. In a living cell, an effector may consist of a single enzyme, a set of enzymes facilitating synthesis of a specific product, or an even more advanced machine-like structure. In the regulatory mechanisms of organism, effectors are often specialized tissues or organs. A typical type of effector mechanism is involved in transcription and translation processes.

The role of the effector is to stabilize the controlled process. Its structure may address the *what?* and (possibly) *where?* questions associated with any biologically active entity, but it primarily relates to the *how?* question by determining the mechanism applied for a given task, as requested by the receptor. We can state that the genetic code (i.e., nucleotide sequence) describes the primary structure of receptor, effector, and transfer proteins. A regulatory loop (negative feedback loop) is a self-contained functional unit which performs a specific task in an automated manner.

Self-organization mechanisms determine the location of biological structures both in individual cells and in organisms—thus, they address the *where?* question. Structures built according to the genetic blueprint and interacting in specific ways may spontaneously generate complexes, associates, and as well a set of cells which may then recognize one another through appropriate receptor systems. Examples of self-organization include spontaneous formation of the cellular membrane from enzymatically synthesized phospholipids (which arrange themselves into planar micelles in the presence of water) and the mutual recognition of tissue cells (Figs. 3.27, 3.28, 3.29, 3.30, 3.31, 3.32, and 3.33).

The *where?* question is particularly important in the development and maturation of organisms. A crucial issue is how to create spatial points of reference in a developing embryo, enabling precise distribution of organs and guiding the development process as a whole. The most frequently applied strategy is to divide the embryo into specific parts, each with a different biological "address," and to apply a separate control process to each part. Such division occurs in stages and is guided by sequentially activated gene packets, according to a predetermined genetic algorithm.

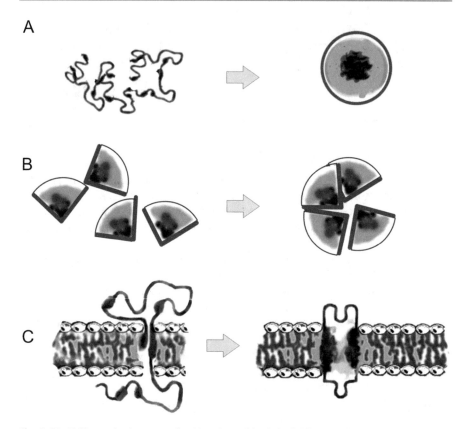

**Fig. 3.28** Self-organization example: (**a**) polypeptide chain folding, (**b**) formation of quaternary protein structure, and (**c**) integration of proteins in the cellular membrane

The ability to assign permanent "addresses" to individual components of the organism is a result of cell differentiation. Following spatial self-orientation of the embryo (mediated by *hox* genes), each "address" is targeted for signals which either promote or inhibit cell proliferation and further specialization, resulting in development of specific organs.

This strategy is evident in insect embryos, particularly in the oft-studied fruit fly (*Drosophila melanogaster*). It relies on three basic gene packets whose sequential activation results in structural self-orientation of the embryo and progressive development of the organism. These packets are, respectively, called *maternal genes*, *segmentation genes*, and *homeotic genes*.

The fruit fly egg already exhibits discernible polarity. During development the embryo undergoes further segmentation which clearly defines points of reference and enables precise placement of organs. Segmentation can commence once the frontal, rear, ventral, and dorsal areas of the embryo are determined. This process, in turn, relies on mechanisms activated by the mother inside the egg (this is why the relevant gene packed is called *maternal*). Following initial self-determination, it

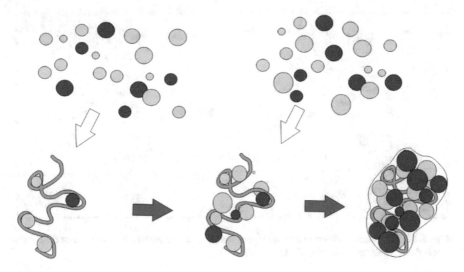

**Fig. 3.29** Simplified view of the self-organization of a ribosome subunit through sequential binding of proteins

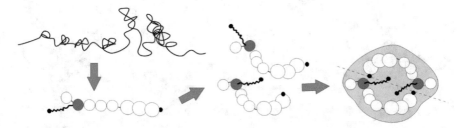

**Fig. 3.30** Simplified view of the formation of fatty acid synthase through self-organization

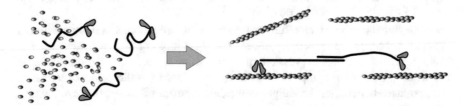

**Fig. 3.31** Simplified view of the self-organization of skeletal muscle—initial stage

becomes possible to delineate boundaries and segments by way of contradictory activity of cells making up each of the preexisting polar regions. This process is mediated by hormones (morphogens) or by direct interaction between adjacent cells. The creation of boundaries is similar to a geopolitical process where two neighboring countries compete to control as much land as possible, ultimately reaching a detente which translates into a territorial border. Differentiated boundary cells

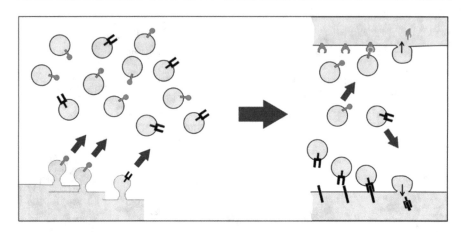

**Fig. 3.32** Distribution of substances encapsulated in vesicles, surrounded by a membrane with integrated markers assisting the self-organization process

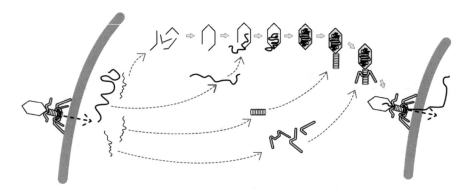

**Fig. 3.33** Self-organization of a phage with the use of components synthesized by a bacterial cell

generate signals which induce further segmentation. This process continues until a suitable precision is reached, under the guidance of *segmentation genes*. Transcription-dependent expression of these genes results in cell differentiation and determines the final purpose of each segment. Once specific points of reference (i.e., segment boundaries) are in place, the development of organs may commence, as specified by *homeotic genes*.

To illustrate the need for this strategy, Fig. 3.34 presents how a blind tailor would go about making a dress. He begins by marking the cloth and then uses these marks to recreate the structure which exists in his mind.

The aim of the example is to visualize the purpose of natural strategies observed in embryonic development (note that this example does not fully reflect the properties of biological processes).

## 3.4    Information Entropy and Mechanisms Assisting Selection

According to the second law of thermodynamics, any isolated system tends to approach its most probable state which is associated with a relative increase in entropy. Regulatory mechanisms can counteract this process but require a source of information. A steady inflow of information is therefore essential for any self-organizing system.

From the viewpoint of information theory, entropy can be described as a measure of ignorance. Regulatory mechanisms which receive signals characterized by high degrees of uncertainty must be able to make informed choices to reduce the overall entropy of the system they control. This property is usually associated with development of information channels. Special structures ought to be exposed within information channels connecting systems of different characters as, for example, linking transcription to translation or enabling transduction of signals through the cellular membrane. Examples of structures which convey highly entropic information are receptor systems associated with blood coagulation and immune responses.

The regulatory mechanism which triggers an immune response relies on relatively simple effectors (complement factor enzymes, phages, and killer cells) coupled to a highly evolved receptor system, represented by specific antibodies and organized set of cells. Compared to such advanced receptors, the structures which register the concentration of a given product (e.g., glucose in blood) are rather primitive.

Advanced receptors enable the immune system to recognize and verify information characterized by high degrees of uncertainty. The system must be able to distinguish a specific antigen among a vast number of structures, each of which may potentially be treated as a signal.

The larger the set of possibilities, the more difficult it is to make a correct choice—hence the need for intricate receptor systems. The development and evolution of such systems increase the likelihood that each input signal will be recognized and classified correctly.

In sequential processes it is usually the initial stage which poses the most problems and requires the most information to complete successfully. It should come as no surprise that the most advanced control loops are those associated with initial stages of biological pathways. The issue may be roughly compared to train travel. When setting out on a journey, we may go to the train station at any moment we wish and then board any train, regardless of its destination. In practice, however, our decision must take into account the specific goal of our journey. The number of decisions required at this preliminary stage is high: we need to decide whether we wish to travel at all, in which direction, on which train, from which platform, and so on. We also have to make sure that the train waiting at the platform is the one we wish to board. In systems devoid of sentience, such questions must be "posed" by specific protein structures, attached to the control loop and usually discarded once their task has been fulfilled. The complexes formed at these initial stages of biological processes are called initiators. Additional structural elements (usually protein-based) which "pose questions" through specific interactions facilitate correct

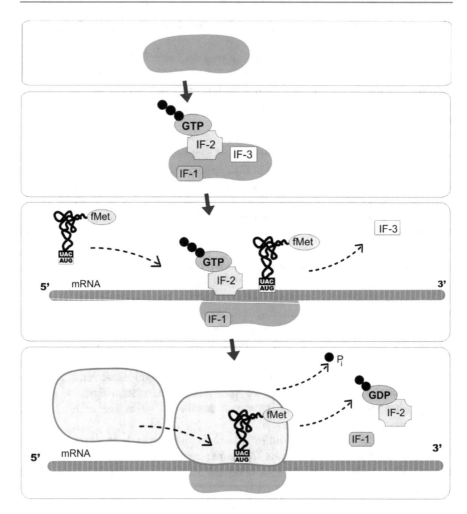

**Fig. 3.35** Simplified view of the creation and disassembling of the initiation complex assisting protein synthesis in prokaryotes

selections among many seemingly random possibilities. Figure 3.35 presents an example of the formation and degradation of initiation complexes in the synthesis of proteins in prokaryotic cell.

## 3.5    Indirect Storage of Genetic Information

While access to energy sources is not a major problem, sources of information are usually far more difficult to manage—hence the universal tendency to limit the scope of direct (genetic) information storage. Reducing the length of genetic code enables efficient packing and enhances the efficiency of operations while at the same time decreasing the likelihood of errors. A classic example of this trend is the progressive evolution of alternative splicing of exon fragments. The number of genes identified in the human genome is lower than the number of distinct proteins by a factor of 4, a difference which can be attributed to alternative splicing. Even though the set of proteins which can be synthesized is comparatively large, genetic information may still be accessed in a straightforward manner as splicing occurs in the course of synthesizing final mRNA chains from their pre-mRNA precursors. This mechanism increases the variety of protein structures without affecting core information storage, i.e., DNA sequences.

The information expressed as a sequence of amino acids in the polypeptide chain is initially contained in the genome; however, the final product of synthesis—the protein itself—may be affected by recombination of exon fragments. Thus, there is no direct correspondence between synthesized proteins and their genetic representation.

### 3.5.1 Self-Organization as a Means of Exploiting Information Associated with the Natural Direction of Spontaneous Processes

In addition to information contained directly in nucleotide sequences, the cell genome also carries "unwritten" rules, rooted in evolutionary experience. Such experience can be explained as a form of functional optimization, resulting from deletion of detrimental and deadweight solutions from genetic memory. This mechanism also applies to information which proves redundant once a simpler solution to a particular problem has been found.

Evolutionary experience may also free the genome from unnecessary ballast by exploiting certain mechanisms by which a reaction may draw information from sources other than the genome itself. This is possible by, e.g., exploiting the natural direction of spontaneous processes. If the lumberjack knows that the sawmill is located by the river, he does not have to carry a map—he simply needs to follow the riverbed. A similar situation may occur while sailing: if the intended direction of travel is consistent with wind direction, all we need to do is set a sail—we do not require knowledge of paddling or navigation.

Such "unwritten" information is a classic example of natural self-organization at work. However, it requires a suitable initial structure, synthesized in accordance with a genetic blueprint.

Self-organization can be likened to stones randomly rolling down a hill and accumulating at its foot. We can expect that such stones will be mostly round in shape, as flat or otherwise uneven stones are not as likely to roll. Clearly, roundness is necessary to exploit the force of gravity as a means of propulsion. Flat stones would instead need to be carried down the hill (which, of course, requires an additional source of energy and information).

As mentioned above, biological self-organization is most frequently associated with protein folding and synthesis of cellular membranes.

The primary structure of the polypeptide chain provides a starting point for the emergence of higher-order structures. The folding process is spontaneous and is, however, largely irrespective of information stored in the genome. While the genetic information is present in the amino acid sequence of the polypeptide chain, folding does not directly rely on it. In order to fold properly, the protein must draw information from its environment. The primary source of such information is the presence of water whose molecules assume a specific structural order in the presence of a hydrophobic residue of the polypeptide and then, by spontaneously reverting to their unordered state, transfer this information back to the polypeptide chain driving its folding.

In terms of energy flow, polypeptide chain folding is a spontaneous process, consistent with the second law of thermodynamics. It is assumed that the final, native form of the protein corresponds to an energy minimum of the protein-water system.

Self-organization of the polypeptide chain is powered primarily by hydrophobic interactions which may be described as a thermodynamically conditioned search for the optimal structure of the chain in the presence of water.

Most researchers believe that only those proteins whose global energy minima correspond to biological activity pass evolutionary selection and become encoded in the genome. Consequently, their secondary and tertiary structure may emerge through spontaneous interaction of the polypeptide chain with its environment.

Although the formation of three-dimensional structures does not directly depend on nucleotide sequences, the cell may nevertheless employ special proteins called chaperones, assisting polypeptide chains in finding their optimal conformation (i.e., their energy minima). The role of chaperones is to prevent aggregation of partly folded chains and promote correct packing of chains by restricting their freedom. They do not directly interfere in folding—instead, their contribution may be treated as a form of genetic interference to self-organization.

Owing to evolutionary selection of polypeptide chains which ensure spontaneous synthesis of the required structures in an aqueous environment (corresponding to active forms of proteins), the genome does not need to directly encode information related to the extremely complex folding process.

Self-organization may also yield more advanced structures consisting of multiple proteins—such as ribosomes and other cellular organelles.

Specific reactivity is revealed by proteins recognizing and binding their specific markers; thus determining localization also occurs in the context of self-organization, although its contribution to this process is often limited. It can be observed, e.g., in intercellular interaction where the participating receptors are sometimes called *topological receptors*.

### 3.5.1.1  Formation of Organized Structures as a Means of Reducing the Necessary Quantity of Information

An organized system is, by definition, more efficient in exploiting information than an unorganized system. The emergence of complex structures through self-organization of genetically programmed components, resulting in improved operational efficiency, can be explained as a form of utilizing information which is not directly contained in the genome. Subunits of protein complexes owe their connectivity to DNA-encoded structural properties, yet their aggregation is a spontaneous process, independent of any genetic representation. It occurs as a consequence of the structural affinity of subunits and does not directly translate into any form of code. This "design concept," concealed in the structure of subunits, expresses itself via their interactions in protein complexes. Advanced complexes act as biological machines and are capable of operating with no need for large quantities of information (compared to individual subunits). Examples of such structures include ribosomes, DNA and RNA polymerases, proteasomes, chaperones, etc.

The amount of required information can be further reduced by restricting selected functions to specific areas of the cell. An example of this process (also called compartmentalization) is the delegation of fatty acid degradation processes to mitochondria, which allows the cell to separate such processes from synthesis of new fatty acid molecules. Conducting both actions in a shared space would require additional regulatory mechanisms and therefore additional genetic code. We should also note the clear division of chromatin present in the nucleus of eukaryotic cells into introns and exons, which appears to play an important role in evolutionary development. In prokaryotes, translation is intimately coupled to transcription, and genes cannot exist as discrete units (maintaining discrete genes would require an unfeasibly large set of additional regulatory mechanisms).

### 3.5.1.2 Reducing the Need for Genetic Information by Substituting Large Sets of Random Events for Directed Processes

A stochastic (directionless) system may fulfill a specific task purely through randomization and selection. The likelihood of achieving the required effect increases in relation to the number of events. Thus, meeting the stated goal (performing a specific action) can result from a trial-and-error approach, given a large enough number of tries. In biological systems, directed processes (requiring information) are frequently replaced by large pools of random actions which can occur with limited input. This model can be compared to operating a machine gun which fires many ($k$) bullets, each with a small but non-negligible probability ($p$) of hitting the target, as opposed to launching a single guided missile which has a very high probability ($p = 1.0$) of impacting the same target. In the former case, the likelihood of a successful hit increases with the number of bullets fired, whereas in the latter case, it depends on the quality of electronic guiding systems. The guided missile is highly efficient (we only need one), but producing and operating it require a vast quantity of information. In contrast, the machine gun is a relatively primitive weapon, yet given a large enough number of tries, it also offers a good chance of scoring a hit.

It should therefore be quite natural to employ stochastic strategies in directionless (i.e., non-sentient) biological systems, where the cost of increasing the number of attempts is far lower than the cost of obtaining additional information.

If the probability of achieving a hit on each attempt is equal to $p$ and all attempts are mutually independent, the overall likelihood of hitting the target ($P$) is expressed as

**Fig. 3.37**  Increasing the likelihood of achieving a biological goal by broadening the distribution of seeds (increased $p$): dandelion seed (wind action). Broadening seed distribution (increased $p$): linden and maple seeds (rotary motion). Broadening seed distribution (increased $p$): greater burdock seeds (adhesion to animal fur)

$$P = 1-(1-p)^k$$

where $p$ is the probability of hitting the target on any given attempt (probability of the elementary event) and $k$ is the number of bullets fired. Figure 3.36 shows increases in $P$ as a consequences of more accurate targeting (increased $p$) and a larger number of attempts (increased $k$).

This problem often emerges in interactions between biological systems and their external environment where the goal is poorly defined (i.e., $p$ is low). Let us consider the odds that a plant seed will encounter favorable ground in which it can germinate. Clearly, the dominant biological strategy is to produce a large number of seeds, increasing the chance that at least one will be successful; however increasing $p$ is also possible—for example, by broadening the dispersal radius with the aid of suitable biological structures (Fig. 3.37).

Intracellular mechanisms usually involve fewer attempts due to the limited space in which they operate. One example is the search for target objects, e.g., the interaction between microtubules and chromosomes. Microtubules form a dynamic system which fluctuates as each individual microtubule grows or shrinks. Their growth depends on the number and activity (by way of forming complexes with GTP) of tubulin molecules, which exhibit GTPase properties. Given a suitable concentration of active (GTP-bound) molecules, microtubules are capable of random growth. The distribution of tubulin is uneven, as newly formed molecules are quickly integrated into growing chains. Areas characterized by rapid growth of microtubules become devoid of tubulin, which in turn precludes further growth. Only those microtubules which find their targets (by associating with an external object) gain stability. This system is capable of sweeping the cell area and, given a large enough number of steps, locating all chromosomes. We can expect that limiting the search space to the interior of a cell results in relatively high $p$ and therefore the number of tries ($k$) may be lower than in a system which is forced to act in open, unrestricted space (Fig. 3.38).

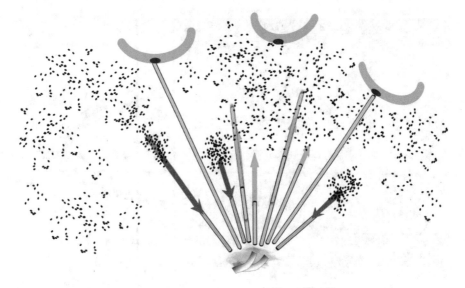

**Fig. 3.38** Spatial search mechanism, as implemented by microtubules

Development of the mitotic spindle. Gray arcs represent chromosomes, while black dots represent tubulin molecules.

Another example of substituting a stochastic process for a directed one is the formation of antibodies. Since the probability that any single antibody will match the given antigen ($p$) is extremely low, the number of randomly generated antibodies ($k$) must be correspondingly high.

Antibody differentiation clearly relies on the "large numbers" strategy, assisted by the "accelerated evolution" mechanism discussed above.

The V exon of both the light and heavy antibody chains emerges through a recombination process which resembles building a house from toy blocks. The randomness of DNA fragments making up each V section, together with the large number of elements which participate in recombination, provides the resulting antibodies with random specificity. Given the great number of synthesized antibodies (high $k$), it is exceedingly likely that any given antigen will be recognized by at least some antibodies ($P = 1.0$) (Fig. 3.39).

By maintaining an enormous, active receptor (consisting of antibodies and specialized cells) where the number of different elements is in the $10^8$–$10^{11}$ range, the immune system can respond to a great variety of potential threats. However, such a strategy requires a vast production line, responsible for replacing lost components and ensuring constant alertness. This is akin to a burglar who carries a huge bunch of keys, hoping that at least one of them will match the lock he is trying to open (Fig. 3.40).

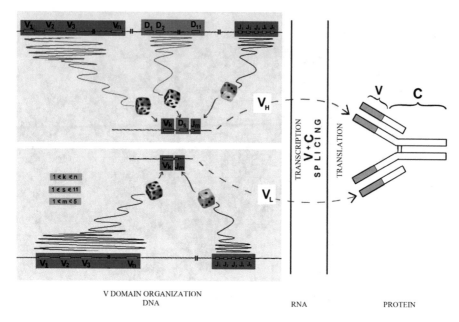

V DOMAIN ORGANIZATION
DNA                                              RNA                          PROTEIN

**Fig. 3.39** Randomness in antibody synthesis

**Fig. 3.40** Symbolic representation of the "large numbers" strategy—different numbers of keys (1 to *n*) and the increasing probability that a given door can be opened by at least one key in the bunch

**1 2 3 4 5 . . . n**

### 3.5.1.3 Exploiting Systemic Solutions as a Means of Restricting Information Requirements

This principle corresponds to an adjustable skeleton key which can be used to pick many locks and therefore replaces a large bunch of individual keys.

Finding a generalized operating principle (by determining commonalities among a large number of individual mechanisms) is a good way to reduce the need for directly encoded information. Such generalizations enable the organism to apply a single procedure to a wide variety of situations.

One example of this strategy is organic detoxification, i.e., transforming toxic compounds into their polar metabolites which are far less active in the organism and can easily be excreted. Here, instead of a large number of specific detoxification processes (as in the case of the immune system, where each antigen has a specific

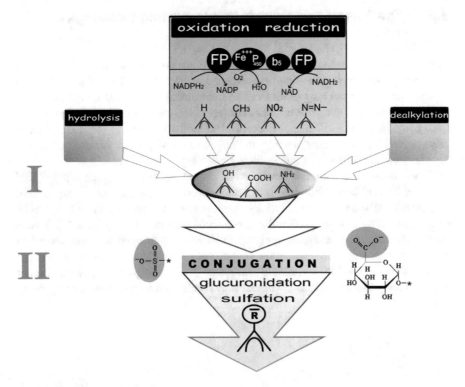

**Fig. 3.41** Detoxification mechanisms active in the endoplasmic reticulum of hepatocytes

associated antibody), a relatively small group of enzymes may effectively detoxify the organism by applying the common principle of reducing the toxicity of dangerous compounds by increasing their polarity.

Figure 3.41 presents detoxification as applied to aromatic compounds. The enzymatic systems integrated in the endoplasmic reticulum conduct oxidation-reduction processes dominated by monooxygenase activity. Other processes (including hydrolysis, deamination, etc.) occur simultaneously, resulting in synthesis of reactive, polar metabolites which contain R–OH, R–NH$_2$, and R–COOH groups.

Application of systemic principle—increased polarity results in decreased toxicity.

At the next stage of the detoxification process, such groups can readily associate with other highly polar compounds such as glucuronic acid, sulfate group, taurine, etc., yielding polar, water-soluble derivatives which are easily captured by the kidneys and excreted in urine.

## 3.6    The Role of Information in Interpreting Pathological Events

Intuition suggests that controlled (regulated) processes fall within the domain of physiology, while processes that have escaped control should be considered pathological. Loss of control may result from an interruption in a regulatory circuit. Thus, information is the single most important criterion separating physiological processes from pathological ones. Diabetes is caused by insulin deficiencies (or insulin immunity), genetic defects may emerge in the absence of certain effector enzymes, neoplasms occur when cells ignore biological "stop" signals, etc.

When discussing immune reactions, it should be noted that an exaggerated response to certain stimuli may prove just as detrimental as a complete lack of response. Similar hypersensitivity may also involve, e.g., nitric oxide, which can be overproduced by LPS or TNF and cause pathological reactions (Fig. 3.42). Pathological deficiencies sometimes apply to vitamins (which need to be delivered in food) and other substances which the organism expects to have available as a result of evolutionary conditioning. Finally, instances of poisoning (where the function of enzymes and other proteins is disrupted) may result from a breakdown of regulatory loops.

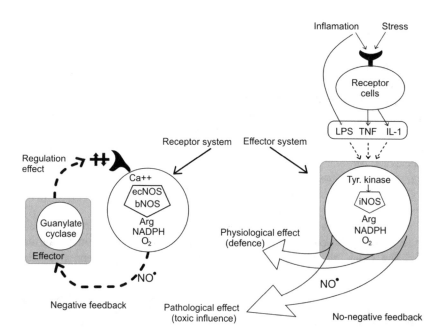

**Fig. 3.42** Physiological (controlled) and pathological (uncontrolled) effects of nitric oxide—efficient regulation as a requirement in physiological processes

## 3.7 Hypothesis

### 3.7.1 Protein Folding Simulation Hypothesis: Early-Stage Intermediate Form

The verification of the correctness of the research presented above, as well as its conclusions, may be verified by attempting to simulate biological phenomena in silico (this is a counterpart of the terms in vivo and in vitro which describe research environments—a less popular equivalent is the term *in computro*). Suitable computer programs, reflecting the properties of real-life biological systems, may serve as an important tool for verifying scientific hypotheses.

The folding of polypeptide chains into their native three-dimensional forms is a prerequisite of biological activity. The spatial structure of a protein molecule determines its interaction with other molecules, substrates (in the case of enzymes), ligands (in the case of coenzymes), and prosthetic groups, and its structural lability (i.e., inherent instability, understood as an effect which facilitates biological function—note that the structure of the protein is not rigid and may undergo changes) in addition to any allosteric properties it may exhibit. Under normal conditions polypeptide chain folding is a directed process which yields a specific, predetermined structure.

While in silico methods are quite effective in genomics, calculation of tertiary and quaternary structures has so far proven elusive. We should note that protein folding is not the only area where computational techniques may be exploited in support of biological research. Equally important is the ability to determine the function of a given protein by identifying its ligand binding sites, active sites (in the case of enzymes), or the ability to form complexes with other proteins.

Simulating three-dimensional structures becomes particularly important in the context of genomics-based successful identification of genes of proteins unknown in biochemistry. An effective means of simulating protein structures would enable us to determine their role in biological systems.

Thus far, the biennial Critical Assessment of Structure Prediction (CASP) experiment, organized for the tenth time in 2012, has not produced significant progress, despite the involvement of key research centers from around the world (see http://predictioncenter.org).

Theoretical models assessed within the scope of CASP can be divided into two groups: Boltzmann and Darwinian approaches. The former group assumes that the polypeptide chain changes its conformation along the energy gradient according to the so-called thermodynamic hypothesis (which states that protein folding is simply an ongoing "quest" to reach an energy minimum). The Darwinian approach bases on evolutionary criteria, claiming that proteins have evolved to attain their observed structure and function. Thus, Darwinists focus on structural comparisons, especially those involving homologous proteins (i.e., proteins which share common ancestry). Research teams which apply the thermodynamic hypothesis seek global optimization methods as a means of performing structural assessment. In contrast, scientists who follow the Darwinian approach query protein databases in search for similar structures and sequences, trying to determine the evolutionary proximity of various proteins. If an existing protein is found to be structurally similar to a new amino acid sequence (for which only the primary structure is known), the folding properties of this new sequence can be confidently predicted on the basis of the available templates.

The model presented below (as a hypothesis) bases on accurate recreation of the folding process, contrary to methods which rely on guessing the structure of a given amino acid chain. In addition to finding out the spatial conformation of a known sequence of amino acids, a useful research tool should also propose a generalized model of the folding process itself, enabling researchers to perform in silico experiments by interfering with the described mechanism (e.g., by introducing mutations and determining their impact on the biological function of the protein). Such techniques would be particularly useful in drug research where the aim is to design a drug with specific properties.

Before presenting the model, however, we first need to introduce some general concepts relating to stages which precede translation.

The foundation of biological systems is the flow of information—from genetic code to a three-dimensional protein structure, capable of performing a specific function:

$$DNA => RNA => Protein$$

Or, to be more exact:

$$DNA => mRNA => AA => 3D => Biological\ function$$

AA indicates the primary structure of the protein, while 3D stands for its spatial (three-dimensional) structure.

Numerical (stochastic) analysis of DNA nucleotide sequences enables us to locate genes, i.e., fragments which are subject to transcription and expression in the form of mRNA chains. The sequence of nucleotides in each mRNA chain determines the sequence of amino acids (AA) which make up the protein molecule. This sequence, in turn, determines the structural (3D) form of the protein and therefore its biological function.

While modern in silico sequencing techniques (including gene identification) appear sufficiently reliable, and the translation process is mostly deterministic (we know which sequences correspond to each amino acid), transforming an amino acid sequence into a three-dimensional protein structure remains an exceedingly difficult problem.

Proteins attain their 3D structure through self-organization. From the viewpoint of energy management, this process conforms to the so-called thermodynamic hypothesis which states that a folding polypeptide chain undergoes structural changes which lower its potential energy in search for an energy minimum.

In light of the stated requirement for accurate recreation of experimental conditions, we need to consider the fact that folding occurs in steps. The presented information flow model can be extended with intermediate structural forms:

$$\text{DNA} = > \text{mRNA} = > \text{AA} = > I_1 = > I_2 = > \dots I_n = > 3\text{-D} =$$
$$> \text{Biological function}$$

$I_1–I_n$ indicate an arbitrary number of intermediate stages.

A slightly more specific model which assumes two distinct intermediate stages called ES (*early stage*) and LS (*late stage*) will be discussed later on:

$$\text{DNA} = > \text{mRNA} = > \text{AA} = > \text{ES} = > \text{LS} = > 3\ \text{D} = > \text{Biological function}$$

In order to determine how ES and LS are generated, let us consider the first stage in the information pathway. Genetic information is encoded by a four-letter alphabet where each letter corresponds to a nucleotide. In contrast, the amino acid "alphabet" consists of 20 separate letters—one for each amino acid.

According to information theory, the quantity of information carried by a single nucleotide is 2 bits ($-\log_2(1/4)$), whereas the quantity of information needed to unambiguously select one amino acid from 20 is 4.23 bits ($-\log_2(1/20)$). Clearly, a single nucleotide cannot encode an amino acid. If we repeat this reasoning for nucleotide pairs, we will conclude that two nucleotides are still insufficient (they only carry 4 bits of information—less than the required 4.23 bits). Thus, the minimum number of nucleotides needed to encode a single amino acid is 3, even though such a sequence carries more information that is strictly needed (specifically, 6 bits). This excess information capacity explains the redundancy of genetic code.

Let us now take this elementary link between probability and the quantity of information required in the translation process and apply it to subsequent stages in the information pathway, according to the central dogma of molecular biology.

If the sequence of amino acids unambiguously determines the three-dimensional structure of the resulting protein, it should contain sufficient information to permit the folding process to take place.

As already noted, one amino acid carries approximately 4.23 bits of information (assuming that all amino acids are equally common). How much information must be provided to describe the conformation of a given amino acid? This property is determined by deriving the so-called conformer, i.e., the value of two dihedral angles which correspond to two degrees of freedom: $\Phi$ (Phi), the angle of rotation about the N–$C_\alpha$ bond, and $\Psi$ (Psi), the angle of rotation around the N–$C_\alpha$ and $C_\alpha$–C', bond where N stands for the amine group nitrogen, while C' stands for the carbonyl group carbon. Each angle may theoretically assume values from the $-180$ to $180°$ range. The combination of both angles determines the conformation of a given amino acid within the polypeptide sequence. All such combinations may be plotted on a planar chart where the vertical (X) axis corresponds to $\Phi$ angles, while the horizontal (Y) axis determines $\Psi$ angles. This chart is called the Ramachandran plot, in honor of its inventor. It spans the entire conformational space, i.e., it covers all possible combinations of $\Phi$ and $\Psi$ angles.

Let us now assume that it is satisfactory to measure each angle to an accuracy of $5°$. The number of possible combinations is equal to $(360/5) * (360/5) = 72 * 72$

**Table 3.3** The quantity of information carried by each amino acid ($I_A$) (column B) in relation to the quantity of information (interpreted as entropy) required to assign its conformation to a specific area on the Ramachandran plot with an accuracy of 5° (column C) or 10° (column D) (correspondingly—$I_5$ and $I_{10}$). Columns E, F, and G present the quantity of information required to pinpoint the location of the peptide bond within a specific cell of the limited conformational subspace (defined in the presented model)

| A | B | C | D | E | F | G |
|---|---|---|---|---|---|---|
| Amino acid | $I_A$ [bit] | $I_5$ [bit] | $I_{10}$ [bit] | $IE_1$ [bit] | $IE_5$ [bit] | $IE_{10}$ [bit] |
| GLY (glycine) | 3.727 | 10.60 | 8.87 | 7.806 | 6.630 | 5.740 |
| ASP (aspartic acid) | 4.121 | 9.68 | 7.81 | 7.073 | 5.950 | 5.016 |
| LEU (leucine) | 3.549 | 8.86 | 7.04 | 6.438 | 5.380 | 4.437 |
| LYS (lysine) | 3.937 | 9.45 | 7.62 | 6.789 | 5.710 | 4.764 |
| ALA (alanine) | 3.661 | 8.86 | 7.00 | 6.409 | 5.419 | 4.462 |
| SER (serine) | 4.060 | 9.69 | 7.81 | 6.975 | 5.785 | 4.857 |
| ASN (asparagine) | 4.494 | 9.90 | 8.11 | 7.267 | 6.126 | 5.186 |
| GLU (glutamic acid) | 3.905 | 9.05 | 7.22 | 6.520 | 5.498 | 4.550 |
| THR (threonine) | 4.107 | 9.49 | 7.66 | 6.720 | 5.502 | 4.579 |
| ARG (arginine) | 4.362 | 9.29 | 7.47 | 6.677 | 5.600 | 4.650 |
| VAL (valine) | 3.868 | 8.89 | 7.06 | 6.233 | 5.057 | 4.108 |
| GLN (glutamine) | 4.684 | 9.16 | 7.36 | 6.676 | 5.607 | 4.667 |
| ILE (isoleucine) | 4.203 | 8.78 | 6.91 | 6.208 | 5.064 | 4.115 |
| PHE (phenylalanine) | 4.679 | 9.37 | 7.52 | 6.617 | 5.466 | 4.528 |
| TYR (tyrosine) | 4.836 | 9.33 | 7.53 | 6.685 | 5.498 | 4.574 |
| PRO (proline) | 4.451 | 8.33 | 6.51 | 6.124 | 4.958 | 4.062 |
| HIS (histidine) | 5.461 | 9.67 | 7.92 | 6.965 | 5.805 | 4.868 |
| CYS (cysteine) | 5.597 | 9.71 | 7.94 | 6.937 | 5.720 | 4.792 |
| MET (methionine) | 5.636 | 8.86 | 7.12 | 6.494 | 5.425 | 4.484 |
| TRP (tryptophan) | 6.091 | 9.10 | 7.38 | 6.581 | 5.444 | 4.512 |

(note, however, that $-180$ and $180°$ angles are in fact equivalent, so we are in fact dealing with 71 * 71 possible structures).

Information theory tells us how many bits are required to unambiguously encode 71 * 71 possible combinations: we need $-\log_2((1/71)(*(1/71)) = 12.29$ bits. Thus, from the point of view of information theory, it is impossible to accurately derive the final values of $\Phi$ and $\Psi$ angles starting with a bare amino acid sequence. However, the presented calculations rely on the incorrect assumption that all amino acids are equally common in protein structures. In fact, the probability of occurrence ($p$) varies from amino acid to amino acid. The Ramachandran plot shows also that some conformations are preferred, while others are excluded due to the specific nature of peptide bonds or their association with high-energy states (which, as already explained, are selected against the folding process). Column B in Table 3.3 presents the quantity of information carried by each amino acid given its actual probability of occurrence in polypeptide chains (these values are derived from a nonredundant protein base—PDB —where multiple data sets obtained by different research institutions for each protein have been merged into a single master set). Columns

C and D indicate the expected quantity of information required to determine the dihedral angles $\Phi$ and $\Psi$ (to an accuracy of 5 or 10°, respectively) for a specific peptide bond, subject to the probability distribution in the Ramachandran plot as well as the probability of occurrence of each amino acid ($p$). The values presented in Table 3.3 reflect entropy which corresponds to the quantity of information (weighted average over the entire conformational space).

The values presented in column B of Table 3.3 (the quantity of information carried by each amino acid and the average quantity of information required to determine its corresponding $\Phi$ and $\Psi$ angles) suggest that simply knowing a raw amino acid sequence does not enable us to accurately model its structure. This is why experimental research indicates the need for intermediate stages in the folding process.

A basic assumption concerning the early stage (ES) says that the structure of the polypeptide chain depends solely on its peptide backbone, with no regard for interactions between side chains. In contrast, the late-stage (LS) structure is dominated by interactions between side chains and may also depend on environmental stimuli. Most polypeptide chains fold in the presence of water (although they may also undergo folding in the apolar environment of a cellular membrane).

Explaining the ES structure requires a suitable introduction.

Let us assume that at early stages in the synthesis of polypeptide chains when interactions between side chains are not yet possible, the conformation of a specific chain is strictly determined by mutual orientation of peptide bond planes being the consequence of the change of dihedral angles $\Phi$ and $\Psi$. This angle (called $V$-angle) is expressed as a value from the 0–180° range. Given these assumptions, a helical structure emerges when the values of $V$-angle is close to 0° because in a helix successive peptide bonds planes share (roughly) the same direction (treated as dipoles). In contrast, values close to $V = 180°$ give rise to a so-called β-*sheet* structure, where the peptide bond planes are directed against each other (note that a peptide bond has polarity and can therefore be assigned a *sense* akin to a vector). As it turns out, the relative angle between two neighboring bonding planes determines the curvature radius of the whole polypeptide chain. The helical structure associated with $V \approx 0$ exhibits a low curvature radius, while the sheet-like structure corresponding to $V \approx 180$ is characterized by straight lines, with a near-infinite curvature radius. Intermediate values of V result in structures which are more open than a helix but not as flat as a sheet. The functional dependency between $V$ and the curvature radius (Fig. 3.43b) for a specific area of the Ramachandran plot representing low-energy bonds is depicted in Fig. 3.43a (note the logarithmic scale for radius values to avoid dealing with very large values for β-sheet).

The approximation function (i.e., the curve which most closely matches an arbitrary set of points) is given as

$$\ln(R) = 3.4 \times 10^{-4} \times V^2 - 2.009 \times 10^{-2} \times V + 0.848 \qquad (3.1)$$

Figure 3.43c depicts the placement of structures which satisfy the above equation on the Ramachandran plot. It appears that certain conformations are preferred in

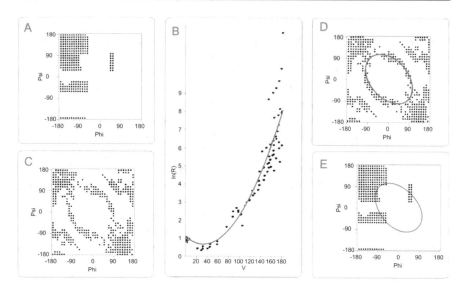

**Fig. 3.43** Elliptical path: (**a**) low-energy area on Ramachandran map, (**b**) relation between V-angle and resultant radius of curvature expressed in logarithmic scale ln(R) with the approximation function, (**c**) the distribution of structures satisfying the relation between V-angle and Ln (R) according to the approximation function, (**d**) elliptical solution graph approximating these points, and (**e**) the ellipse path links all low-energy area on Ramachandran map representing all secondary structural forms

peptide bonds and that there exists a preferred conformational space for the entire peptide backbone (note that the early-stage structure is determined solely by the arrangement of backbone). The elliptical path seen in Fig. 3.43d joins all ordered secondary structures and can be treated as a limited conformation subspace for the early stage (ES) form.

The presented interpretation has one more advantage: if we calculate the quantity of information required to unambiguously select one of the points comprising an ellipse (columns E, F, and G in Table 3.3, for varying degrees of accuracy), we will note that it corresponds to the quantity of information carried by amino acids (column B in Table 3.3).

As a consequence, it seems that the raw sequence of amino acids only contains sufficient information to predict the conformation of the early-stage intermediate form within the subspace presented in Fig. 3.43d and e. The prediction of proper $\Phi_{ES}$ and $\Psi_{ES}$ (ES denotes belonging to the set of points of ellipse path) appears to be possible due to comparable amount of information balancing the information carried by amino acid with the amount of information necessary to predict the ES conformation (limited to ellipse path on Ramachandran map). The ellipse path is treated as limited conformational subspace for ES intermediate in protein folding process.

# Suggested Reading

Abudayyeh OO, Gootenberg JS, Franklin B, Koob J, Kellner MJ, Ladha A, Joung J, Kirchgatterer P, Cox DBT, Zhang F (2019) A cytosine deaminase for programmable single-base RNA editing. Science. 365(6451):382–386. https://doi.org/10.1126/science.aax7063

Ahmad K, Henikoff S (2018) No strand left behind. Science. 361(6409):1311–1312. https://doi.org/10.1126/science.aav0871

Al Mamun AAM, Lombardo M-J, Shee C, Lisewski AM, Gonzalez C, Lin D, Nehring RB, Saint-Ruf C, Gibson JL, Frisch RL, Lichtarge O, Hastings PJ, Rosenberg SM (2012) Identity and function of a large gene network underlying mutagenic repair of DNA breaks. Science 338: 1344–1348

Allegretti M, Zimmerli CE, Rantos V, Wilfling F, Ronchi P, Fung HKH, Lee CW, Hagen W, Turoňová B, Karius K, Börmel M, Zhang X, Müller CW, Schwab Y, Mahamid J, Pfander B, Kosinski J, Beck M (2020) In-cell architecture of the nuclear pore and snapshots of its turnover. Nature. 586(7831):796–800. https://doi.org/10.1038/s41586-020-2670-5

Amadei G, Handford CE, Qiu C, De Jonghe J, Greenfeld H, Tran M, Martin BK, Chen DY, Aguilera-Castrejon A, Hanna JH, Elowitz MB, Hollfelder F, Shendure J, Glover DM, Zernicka-Goetz M (2022) Embryo model completes gastrulation to neurulation and organogenesis. Nature 610(7930):143–153. https://doi.org/10.1038/s41586-022-05246-3

Arroyo M, Trepat X (2019) Embryonic self-fracking. Science 365(6452):442–443. https://doi.org/10.1126/science.aay2860

Ashby WR (1958) An introduction to cybernetics. Chapman & Hall, London

Atabay KD, LoCascio SA, de Hoog T, Reddien PW (2018) Self-organization and progenitor targeting generate stable patterns in planarian regeneration. Science. 360(6387):404–409. https://doi.org/10.1126/science.aap8179

Aushev M, Herbert M (2020) Mitochondrial genome editing gets precise. Nature. 583(7817): 521–522. https://doi.org/10.1038/d41586-020-01974-6

Bailles A, Collinet C, Philippe J-M, Lenne P-F, Munro E, Lecuit T (2019) Genetic induction and mechanochemical propagation of a morphogenetic wave. Nature 572(7770):467–473. https://doi.org/10.1038/s41586-019-1492-9

Barisic D, Stadler MB, Iurlaro M, Schübeler D (2019) Mammalian ISWI and SWI/SNF selectively mediate binding of distinct transcription factors. Nature. 569(7754):136–140. https://doi.org/10.1038/s41586-019-1115-5

Bernstein HD (2012) All clear for ribosome landing. Nature 492:189–191

Bertero A, Brown S, Madrigal P, Osnato A, Ortmann D, Yiangou L, Kadiwala J, Hubner NC, de Los Mozos IR, Sadée C, Lenaerts AS, Nakanoh S, Grandy R, Farnell E, Ule J, Stunnenberg HG, Mendjan S, Vallier L (2018) The SMAD2/3 interactome reveals that TGFβ controls m6A mRNA methylation in pluripotency. Nature. 555(7695):256–259. https://doi.org/10.1038/nature25784

Bobrovskiy I, Hope JM, Ivantsov A, Nettersheim BJ, Hallmann C, Brocks JJ (2018) Ancient steroids establish the Ediacaran fossil Dickinsonia as one of the earliest animals. Science. 361(6408):1246–1249. https://doi.org/10.1126/science.aat7228

Brennan MD, Cheong R, Levchenko A (2012) How information theory handles cell signaling and uncertainty. Science 338:334–335

Brown A, Shao S, Murray J, Hegde RS, Ramakrishnan V (2015) Structural basis for stop codon recognition in eukaryotes. Nature. 524(7566):493–496. https://doi.org/10.1038/nature14896

Bycroft M (2011) Recognition of non-methyl histone marks. Curr Op Struct Biol. 21:761–766

Chambers I, Silva J, Colby D, Nichols J, Nijmeijer B, Robertson M, Vrana J, Jones K, Grotewold L, Smith A (2007) Nanog safeguards pluripotency and mediates germline development. Nature 450:1230–1234

Chin JW (2012) Reprogramming the genetic code. Science 336:428–429

Cramer P, Bushnell DA, Fu J, Gnatt AL, Maier-Davis B, Thompson NE, Burgess RR, Edwards AM, David PR, Korneberg RD (2000) Architecture of RNA polymerase II and implications for the transcription mechanism. Science 288:640–649

Cui Y, Lyu X, Ding L, Ke L, Yang D, Pirouz M, Qi Y, Ong J, Gao G, Du P, Gregory RI (2021) Global miRNA dosage control of embryonic germ layer specification. Nature. 593(7860): 602–606. https://doi.org/10.1038/s41586-021-03524-0

de Latil P (1953) La pensée artificielle. Ed. Gallimard, Paris

Dethoff EA, Chugh J, Mustoe AM, Al-Hashimi HM (2012) Functional complexity and regulation through RNA dynamics. Nature 482:322–330

Dever TE (1999) Translation initiation: adept at adapting TIBS 24:398–403

Downs JA, Nussenzweig MC, Nussenzweig A (2007) Chromatin dynamics and the preservation of genetic information. Nature 447:951–958

Emerson DJ, Zhao PA, Cook AL, Barnett RJ, Klein KN, Saulebekova D, Ge C, Zhou L, Simandi Z, Minsk MK, Titus KR, Wang W, Gong W, Zhang D, Venev SV, Gibcus JH, Yang H, Sasaki T, Kanemaki MT, Yue F, Dekker J, Chen C-L, Gilbert DM, Phillips-Cremins JE (2022) Cohesin-mediated loop anchors confine the locations of human replication origins. Nature 606(7915): 812–819. https://doi.org/10.1038/s41586-022-04803-0

Fedoroff NV (2012) Transposable elements, epigenetics and genome evolution. Science 338:758–767

Filbin MG, Tirosh I, Hovestadt V, Shaw ML, Escalante LE, Mathewson ND, Neftel C, Frank N, Pelton K, Hebert CM, Haberler C, Yizhak K, Gojo J, Egervari K, Mount C, van Galen P, Bonal DM, Nguyen QD, Beck A, Sinai C, Czech T, Dorfer C, Goumnerova L, Lavarino C, Carcaboso AM, Mora J, Mylvaganam R, Luo CC, Peyrl A, Popović M, Azizi A, Batchelor TT, Frosch MP, Martinez-Lage M, Kieran MW, Bandopadhayay P, Beroukhim R, Fritsch G, Getz G, Rozenblatt-Rosen O, Wucherpfennig KW, Louis DN, Monje M, Slavc I, Ligon KL, Golub TR, Regev A, Bernstein BE, Suvà ML (2018) Developmental and oncogenic programs in H3K27M gliomas dissected by single-cell RNA-seq. Science. 360(6386):331–335. https://doi.org/10.1126/science.aao4750

Finn EH, Misteli T (2019) Molecular basis and biological function of variability in spatial genome organization. Science 365(6457):eaaw9498. https://doi.org/10.1126/science.aaw9498

Flavahan WA, Drier Y, Johnstone SE, Hemming ML, Tarjan DR, Hegazi E, Shareef SJ, Javed NM, Raut CP, Eschle BK, Gokhale PC, Hornick JL, Sicinska ET, Demetri GD, Bernstein BE (2019) Altered chromosomal topology drives oncogenic programs in SDH-deficient GISTs. Nature 575(7781):229–233. https://doi.org/10.1038/s41586-019-1668-3

Frye M, Harada BT, Behm M, He C (2018) RNA modifications modulate gene expression during development. Science. 361(6409):1346–1349. https://doi.org/10.1126/science.aau1646

Furlong EEM, Levine M (2018) Developmental enhancers and chromosome topology. Science. 361(6409):1341–1345. https://doi.org/10.1126/science.aau0320

Gabriele M, Brandão HB, Grosse-Holz S, Jha A, Dailey GM, Cattoglio C, Hsieh T-HS, Mirny L, Zechner C, Hansen AS (2022) Dynamics of CTCF- and cohesin-mediated chromatin looping revealed by live-cell imaging. Science 376(6592):496–501. https://doi.org/10.1126/science.abn6583

Ganesan A, Zhang J (2012) How cells process information: quantification of spatiotemporal signaling dynamics. Protein Sci 21:918–928

Gkikopoulos T, Schofield P, Singh V, Pinskaya M, Mellor J, Smolle M, Workman JL, Barton GJ, Owen-Hughes T (2011) A role for Snf2-related nucleosome-splicing enzymes in genome-wide nucleosome organization. Science 333:1758–1760

Graf T, Enver T (2009) Forcing cells to change lineages. Nature 462:587–592

Guo YE, Manteiga JC, Henninger JE, Sabari BR, Dall'Agnese A, Hannett NM, Spille JH, Afeyan LK, Zamudio AV, Shrinivas K, Abraham BJ, Boija A, Decker TM, Rimel JK, Fant CB, Lee TI, Cisse II, Sharp PA, Taatjes DJ, Young RA (2019) Pol II phosphorylation regulates a switch between transcriptional and splicing condensates. Nature. 572(7770):543–548. https://doi.org/10.1038/s41586-019-1464-0

Hasty J, McMillen D, Collins JJ (2002) Engineered gene circuits. Nature 420:224–230

He S, Wu Z, Tian Y, Yu Z, Yu J, Wang X, Li J, Liu B, Xu Y (2020) Structure of nucleosome-bound human BAF complex. Science 367(6480):875–881. https://doi.org/10.1126/science.aaz9761

Hoyt SJ, Storer JM, Hartley GA, Grady PGS, Gershman A, de Lima LG, Limouse C, Halabian R, Wojenski L, Rodriguez M, Altemose N, Rhie A, Core LJ, Gerton JL, Makalowski W, Olson D, Rosen J, Smit AFA, Straight AF, Vollger MR, Wheeler TJ, Schatz MC, Eichler EE, Phillippy AM, Timp W, Miga KH, O'Neill RJ (2022) From telomere to telomere: The transcriptional and epigenetic state of human repeat elements. Science 376(6588):eabk3112. https://doi.org/10.1126/science.abk3112

Jin X, Demere Z, Nair K, Ali A, Ferraro GB, Natoli T, Deik A, Petronio L, Tang AA, Zhu C, Wang L, Rosenberg D, Mangena V, Roth J, Chung K, Jain RK, Clish CB, Van der Heiden MG, Golub TR (2020) A metastasis map of human cancer cell lines. Nature 588(7837):331–336. https://doi.org/10.1038/s41586-020-2969-2

Khorasanizadeh S (2011) Recognition of methylated histone new twists and variations. Curr Op Struct Biol. 21:744–749

Kornberg RD (1996) RNA polymerase II transcription control TIBS 21:325–326

Kosak ST, Groudine M (2004) Gene order and dynamic domains. Science 306:644–650

Ku C, Nelson-Sathi S, Roettger M, Sousa FL, Lockhart PJ, Bryant D, Hazkani-Covo E, McInerney JO, Landan G, Martin WF (2015) Endosymbiotic origin and differential loss of eukaryotic genes. Nature 524(7566):427–432. https://doi.org/10.1038/nature14963

Ku C, Nelson-Sathi S, Roettger M, Sousa FL, Lockhart PJ, Bryant D, Hazkani-Covo E, McInerney JO, Landan G, Martin WF (2015) Endosymbiotic origin and differentia loss of eukaryotic genes. Nature 524(7566):427–432. https://doi.org/10.1038/nature14963

Kyprianou C, Christodoulou N, Hamilton RS, Nahaboo W, Boomgaard DS, Amadei G, Migeotte I, Zernicka-Goetz M (2020) Basement membrane remodelling regulates mouse embryogenesis. Nature 582(7811):253–258. https://doi.org/10.1038/s41586-020-2264-2

Lee JT (2012) Epigenetic regulation by long noncoding RNAs. Science 338:1435–1439

Leschziner AE (2011) Electron microscopy studies on nucleosome remodelers. Curr Op. Struct Biol. 21:709–718

Li H, Janssens J, De Waegeneer M, Kolluru SS, Davie K, Gardeux V, Saelens W, FPA D, Brbić M, Spanier K, Leskovec J, CN ML, Xie Q, Jones RC, Brueckner K, Shim J, Tattikota SG, Schnorrer F, Rust K, Nystul TG, Carvalho-Santos Z, Ribeiro C, Pal S, Mahadevaraju S, Przytycka TM, Allen AM, Goodwin SF, Berry CW, Fuller MT, White-Cooper H, Matunis EL, DiNardo S, Galenza A, O'Brien LE, Dow JAT, FCA Consortium§, Jasper H, Oliver B, Perrimon N, Deplancke B, Quake SR, Luo L, Aerts S, Agarwal D, Ahmed-Braimah Y, Arbeitman M, Ariss MM, Augsburger J, Ayush K, Baker CC, Banisch T, Birker K, Bodmer R, Bolival B, Brantley SE, Brill JA, Brown NC, Buehner NA, Cai XT, Cardoso-Figueiredo R, Casares F, Chang A, Clandinin TR, Crasta S, Desplan C, Detweiler AM, Dhakan DB, Donà E, Engert S, Floc'hlay S, George N, González-Segarra AJ, Groves AK, Gumbin S, Guo Y, Harris DE, Heifetz Y, Holtz SL, Horns F, Hudry B, Hung R-J, Jan YN, Jaszczak JS, GSXE J, Karkanias J, Karr TL, Katheder NS, Kezos J, Kim AA, Kim SK, Kockel L, Konstantinides N, Kornberg TB, Krause HM, Labott AT, Laturney M, Lehmann R, Leinwand S, Li J, JSS L, Li K, Li K, Li L, Li T, Litovchenko M, Liu H-H, Liu Y, Lu T-C, Manning J, Mase A, Matera-Vatnick M, Matias NR, McDonough-Goldstein CE, McGeever A, McLachlan AD, Moreno-Roman P, Neff N, Neville M, Ngo S, Nielsen T, O'Brien CE, Osumi-Sutherland D, Özel MN, Papatheodorou I, Petkovic M, Pilgrim C, Pisco AO, Reisenman C, Sanders EN, Dos Santos G, Scott K, Sherlekar A, Shiu P, Sims D, Sit RV, Slaidina M, Smith HE, Sterne G, Su Y-H, Sutton D, Tamayo M, Tan M, Tastekin I, Treiber C, Vacek D, Vogler G, Waddell S, Wang W, Wilson RI, Wolfner MF, Yiu-Wong Y-CE, Xie A, Xu J, Yamamoto S, Yan J, Yao Z, Yoda K, Zhu R, Zinzen RP (2022) Fly Cell Atlas: a single-nucleus transcriptomic atlas of the adult fruit fly. Science 375(6584):eabk2432. https://doi.org/10.1126/science.abk2432

Lickwar CR, Mueller F, Hanlon SE, McNally JG, Lieb JD (2012) Genome-wide protein-DNA binding dynamics suggest a molecular clutch for transcription factor function. Nature 484:251–255

Lim CJ, Barbour AT, Zaug AJ, Goodrich KJ, McKay AE, Wuttke DS, Cech TR (2020) The structure of human CST reveals a decameric assembly bound to telomeric DNA. Science 368(6495):1081–1085. https://doi.org/10.1126/science.aaz9649

Lin X, Liu Y, Liu S, Zhu X, Wu L, Zhu Y, Zhao D, Xu X, Chemparathy A, Wang H, Cao Y, Nakamura M, Noordermeer JN, La Russa M, Wong WH, Zhao K, Qi LS (2022) Nested epistasis enhancer networks for robust genome regulation. Science. 377(6610):1077–1085. https://doi.org/10.1126/science.abk3512

Lister R, Pelizzola M, Kida YS, Hawkins RD, Nery JR, Hon G, Antosiewicz-Bourget J, O'Malley R, Castanon R, Klugman S, Downes M, Yu R, Stewart R, Ren B, Thomson JA, Evans RM, Ecker JR (2011) Hotspots of aberrant epigenomic reprogramming in human induced pluripotent stem cells. Nature 471:68–73

Lombardi PM, Cole KE, Dowling DP, Christianson DW (2011) Structure mechanism and inhibition of histone deacetylases and related metalloenzymes. Curr Op Struct Biol 21:735–743

Luo C, Hajkova P, Ecker JR (2018) Dynamic DNA methylation: in the right place at the right time. Science 361(6409):1336–1340. https://doi.org/10.1126/science.aat6806

Maeshima K (2022) A phase transition for chromosome transmission when cells divide. Nature. 609(7925):35–36. https://doi.org/10.1038/d41586-022-01925-3

Mayor R (2019) Cell fate decisions during development. Science 364(6444):937–938. https://doi.org/10.1126/science.aax7917

Mishra SK, Ammon T, Popowicz GM, Krajewski M, Nagel RJ, Ares M, Holak TA, Jentsch S (2011) Role of ubiquitin-like protein Hub1 in splice-site usage and alternative splicing. Nature 474:173–178

Mitchell L, Chang G, Horton NC, Kercher MA, Pace HC, Schumacher MA, Brennan R, Lu GP (1996) Crystal structure of the lactose operon repressor and its complexes with DNA and inducer. Science 271:1247–1254

Mitter M, Gasser C, Takacs Z, Langer CCH, Tang W, Jessberger G, Beales CT, Neuner E, Ameres SL, Peters JM, Goloborodko A, Micura R, Gerlich DW (2020) Conformation of sister chromatids in the replicated human genome. Nature 586(7827):139–144. https://doi.org/10.1038/s41586-020-2744-4

Mitter M, Takacs Z, Köcher T, Micura R, Langer CCH, Gerlich DW (2022) Sister chromatid-sensitive Hi-C to map the conformation of replicated genomes. Nat Protoc 17(6):1486–1517. https://doi.org/10.1038/s41596-022-00687-6

Moore L, Leongamornlert D, Coorens THH, Sanders MA, Ellis P, Dentro SC, Dawson KJ, Butler T, Rahbari R, Mitchell TJ, Maura F, Nangalia J, Tarpey PS, Brunner SF, Lee-Six H, Hooks Y, Moody S, Mahbubani KT, Jimenez-Linan M, Brosens JJ, Iacobuzio-Donahue CA, Martincorena I, Saeb-Parsy K, Campbell PJ, Stratton MR (2020) The mutational landscape of normal human endometrial epithelium. Nature. 580(7805):640–646. https://doi.org/10.1038/s41586-020-2214-z

Morelli LG, Uriu K, Ares S, Oates AC (2012) Computational approaches to developmental pattering. Science 336:187–191

Moris N, Anlas K, van den Brink SC, Alemany A, Schröder J, Ghimire S, Balayo T, van Oudenaarden A, Arias AM (2020) An in vitro model of early anteroposterior organization during human development. Nature 582(7812):410–415. https://doi.org/10.1038/s41586-020-2383-9

Munsky B, Neuert G, van Oudenaarden A (2012) Using gene expression noise to understand gene regulation. Science 336:183–187

Necsulea A, Soumillon M, Warnefors M, Liechti A, Daish T, Zeller U, Baker JC, Grützner F, Kaessmann H (2014) The evolution of lncRNA repertoires and expression patterns in tetrapods. Nature. 505(7485):635–640. https://doi.org/10.1038/nature12943

Nelms B, Walbot V (2019) Defining the developmental program leading to meiosis in maize. Science. 364(6435):52–56. https://doi.org/10.1126/science.aav6428

Netzer WJ, Hartl FU (1998) Protein folding in the cytosol: chaperonin-dependent and independent mechanism. TIBS 23:68–73

Nicholas CR, Kriegstein AR (2010) Cell reprogramming gets direct. Nature 463:1031–1032

Nurk S, Koren S, Rhie A, Rautiainen M, Bzikadze AV, Mikheenko A, Vollger MR, Altemose N, Uralsky L, Gershman A, Aganezov S, Hoyt SJ, Diekhans M, Logsdon GA, Alonge M, Antonarakis SE, Borchers M, Bouffard GG, Brooks SY, Caldas GV, Chen N-C, Cheng H, Chin C-S, Chow W, de Lima LG, Dishuck PC, Durbin R, Dvorkina T, Fiddes IT, Formenti G, Fulton RS, Fungtammasan A, Garrison E, Grady PGS, Graves-Lindsay TA, Hall IM, Hansen NF, Hartley GA, Haukness M, Howe K, Hunkapiller MW, Jain C, Jain M, Jarvis ED, Kerpedjiev P, Kirsche M, Kolmogorov M, Korlach J, Kremitzki M, Li H, Maduro VV, Marschall T, McCartney AM, McDaniel J, Miller DE, Mullikin JC, Myers EW, Olson ND, Paten B, Peluso P, Pevzner PA, Porubsky D, Potapova T, Rogaev EI, Rosenfeld JA, Salzberg SL, Schneider VA, Sedlazeck FJ, Shafin K, Shew CJ, Shumate A, Sims Y, Smit AFA, Soto DC, Sović I, Storer JM, Streets A, Sullivan BA, Thibaud-Nissen F, Torrance J, Wagner J, Walenz BP, Wenger A, Wood JMD, Xiao C, Yan SM, Young AC, Zarate S, Surti U, McCoy RC, Dennis MY, Alexandrov IA, Gerton JL, O'Neill RJ, Timp W, Zook JM, Schatz MC, Eichler EE, Miga KH, Phillippy AM (2022) The complete sequence of a human genome. Science 376(6588):44–53. https://doi.org/10.1126/science.abj6987

Onuchic V, Lurie E, Carrero I, Pawliczek P, Patel RY, Rozowsky J, Galeev T, Huang Z, Altshuler RC, Zhang Z, Harris RA, Coarfa C, Ashmore L, Bertol JW, Fakhouri WD, Yu F, Kellis M, Gerstein M, Milosavljevic A (2018) Allele-specific epigenome maps reveal sequence-dependent stochastic switching at regulatory loci. Science. 361(6409):eaar3146. https://doi.org/10.1126/science.aar3146

Pawson T, Nash P (2000) Protein-protein interactions define specificity in signal transduction. Genes Dev 14:1027–1047

Pennisi E (2012) Gene duplication's role in evolution gets richer, more complex. Science 338:316–317

Petryk N, Dalby M, Wenger A, Stromme CB, Strandsby A, Andersson R, Groth A (2018) MCM2 promotes symmetric inheritance of modified histones during DNA replication. Science. 361(6409):1389–1392. https://doi.org/10.1126/science.aau0294

Pham P, Bransteitter R, Petruska J, Goodman MF (2003) Progressive AID-catalysed cytosine deamination on single-stranded DNA simulates somatic hypermutation. Nature 424:103–107

Pollard TD, Earnshaw W (2002) Cell biology. Sounders Philadelphia, London

Przeworski M (2005) Motivating hotspots. Science 310:247–248

Rodríguez-Nuevo A, Torres-Sanchez A, Duran JM, De Guirior C, Martínez-Zamora MA, Böke E (2022) Oocytes maintain ROS-free mitochondrial metabolism by suppressing complex I. Nature 607(7920):756–761. https://doi.org/10.1038/s41586-022-04979-5

Rouskin S, Zubradt M, Washietl S, Kellis M, Weissman JS (2014) Genome-wide probing of RNA structure reveals active unfolding of mRNA structures in vivo. Nature 505(7485):701–705. https://doi.org/10.1038/nature12894

Schwarzer W, Abdennur N, Goloborodko A et al (2017) Two independent modes of chromatin organization revealed by cohesin removal. Nature 551:51–56. https://doi.org/10.1038/nature24281

Sekne Z, Ghanim GE, van Roon AM, Nguyen THD (2022) Structural basis of human telomerase recruitment by TPP1-POT1. Science 375(6585):1173–1176. https://doi.org/10.1126/science.abn6840

Shahbazi MN, Siggia ED, Zernicka-Goetz M (2019) Self-organization of stem cells into embryos: a window on early mammalian development. Science 364(6444):948–951. https://doi.org/10.1126/science.aax0164

Sheth R, Marcon L, Bastida MF, Junco M, Quintana L, Dahn R, Kmita M, Sharpe J, Ros MA (2012) Hox genes regulate digit patterning by controlling the wavelength of a Turing-type mechanism. Science 338:1476–1480

Shukla A, Yan J, Pagano DJ, Dodson AE, Fei Y, Gorham J, Seidman JG, Wickens M, Kennedy S (2020) poly(UG)-tailed RNAs in genome protection and epigenetic inheritance. Nature. 582(7811):283–288. https://doi.org/10.1038/s41586-020-2323-8

Solé R (2020) Using information theory to decode network coevolution. Science 368(6497): 1315–1316. https://doi.org/10.1126/science.abc6344

Tan L, Xing D, Chang CH, Li H, Xie XS (2018) Three-dimensional genome structures of single diploid human cells. Science 361(6405):924–928. https://doi.org/10.1126/science.aat5641

Timsit Y (1999) DNA Structure and polymerase fidelity. J Mol Biol 293:835–853

Upadhyay AK, Horton JR, Zhang X, Cheng X (2011) Coordinated methyl-lysine erasure: structural and functional linkage of a Jumonji demethylase domain and a reader domain. Curr Op Struct Biol. 21:750–760

Vierbuchen T, Ostermeier A, Pang ZP, Kokubu Y, Südhof TC, Wernig M (2010) Direct conversion of fibroblast to functional neurons by defined factors. Nature 463:1035–1041

Wan Y, Qu K, Zhang QC, Flynn RA, Manor O, Ouyang Z, Zhang J, Spitale RC, Snyder MP, Segal E, Chang HY (2014) Landscape and variation of RNA secondary structure across the human transcriptome. Nature. 505(7485):706–709. https://doi.org/10.1038/nature12946

Vollger MR, Guitart X, Dishuck PC, Mercuri L, Harvey WT, Gershman A, Diekhans M, Sulovari A, Munson KM, Lewis AP, Hoekzema K, Porubsky D, Li R, Nurk S, Koren S, Miga KH, Phillippy AM, Timp W, Ventura M, Eichler EE (2022) Segmental duplications and their variation in a complete human genome. Science 376(6588):eabj6965. https://doi.org/10. 1126/science.abj6965

Wan Y, Qu K, Zhang QC, Flynn RA, Manor O, Ouyang Z, Zhang J, Spitale RC, Snyder MP, Segal E, Chang HY (2014) Landscape and variation of RNA secondary structure across the human transcriptome. Nature 505(7485):706–709. https://doi.org/10.1038/nature12946

Wilkinson ME, Fica SM, Galej WP, Norman CM, Newman AJ, Nagai K (2017) Postcatalytic spliceosome structure reveals mechanism of 3'-splice site selection. Science. 358(6368): 1283–1288. https://doi.org/10.1126/science.aar3729

Yadav T, Quivy JP, Almouzni G (2018) Chromatin plasticity: a versatile landscape that underlies cell fate and identity. Science 361(6409):1332–1336. https://doi.org/10.1126/science.aat8950

Yang H, Luan Y, Liu T, Lee HJ, Fang L, Wang Y, Wang X, Zhang B, Jin Q, Ang KC, Xing X, Wang J, Xu J, Song F, Sriranga I, Khunsriraksakul C, Salameh LD, Choudhary MNK, Topczewski J, Wang K, Gerhard GS, Hardison RC, Wang T, Cheng KC, Yue F (2020) A map of cis-regulatory elements and 3D genome structures in zebrafish. Nature 588(7837): 337–343. https://doi.org/10.1038/s41586-020-2962-9

Yoshimi A, Lin K-T, Wiseman DH, Rahman MA, Pastore A, Wang B, Lee SC-W, Micol J-B, Zhang XJ, de Botton S, Penard-Lacronique V, Stein EM, Cho H, Miles RE, Inoue D, Albrecht TR, Somervaille TCP, Batta K, Amaral F, Simeoni F, Wilks DP, Cargo C, Intlekofer AM, Levine RL, Dvinge H, Bradley RK, Wagner EJ, Krainer AR, Abdel-Wahab O (2019) Coordinated alterations in RNA splicing and epigenetic regulation drive leukaemogenesis. Nature 574(7777):273–277. https://doi.org/10.1038/s41586-019-1618-0

Yu C, Gan H, Serra-Cardona A, Zhang L, Gan S, Sharma S, Johansson E, Chabes A, Xu RM, Zhang Z (2018) A mechanism for preventing asymmetric histone segregation onto replicating DNA strands. Science. 361(6409):1386–1389. https://doi.org/10.1126/science.aat8849

Zande PV, Hill MS, Wittkopp PJ (2022) Pleiotropic effects of trans-regulatory on fitness and gene expression. Science 377(6601):105–109. https://doi.org/10.1126/science.abj7185

Zhang Y, Zhang X, Ba Z, Liang Z, Dring EW, Hu H, Lou J, Kyritsis N, Zurita J, Shamim MS, Aiden AP, Aiden EL, Alt FW (2019) The fundamental role of chromatin loop extrusion in physiological V(D)J recombination. Nature 573(7775):600–604. https://doi.org/10.1038/ s41586-019-1547-y

# Regulation in Biological Systems

**4**

*Maintaining a steady state in thermodynamically open systems requires regulation.*

## Abstract

Any processes which occur in thermodynamically open systems must be automatically regulated if they are to maintain a steady state. Note that there is an important difference between a steady state and a state of equilibrium, since in the latter case no spontaneity is possible. **The need to maintain a steady state ensuring homeostasis is an essential concern in nature, while negative feedback loop is the fundamental way to ensure that this goal is met.** The regulatory system determines the interdependences between individual cells and the organism, subordinating the former to the latter. In trying to maintain homeostasis, the organism may temporarily upset the steady-state conditions of its component cells, forcing them to perform work for the benefit of the organism. Adopting a systemic approach to the study of regulatory mechanisms explains the mutual dependencies which, taken together, form the foundation of life.

On a cellular level, signals are usually transmitted via changes in concentrations of reaction substrates and products. This simple mechanism is made possible due to limited volume of each cell. Such signaling plays a key role in maintaining homeostasis and ensuring cellular activity. On the level of the organism, signal transmission is performed by hormones and the nervous system. This work addresses the problems of regulation on a systemic level.

## Keywords

Nonequilibrium · Signal transduction · Homeostasis · Biological regulatory programs · The cell and the organism · Control of signaling · Nature of signals

1. Regulation and control—what is the difference?
2. Types of control.
3. What is the importance of the allosteric effect for biological regulation mechanisms?
4. Role and mechanism of action of G proteins.
5. What is the relationship between the organism and the cell in terms of regulation?
6. Why and when does amplification of signals become necessary?
7. When must a regulatory signal be encoded?
8. When must a regulatory signal be actively attenuated?
9. How is function encoded in genes, and how is such information transmitted?
10. What is the cause and logic behind autocrine and paracrine regulation?
11. What role do positive feedback loops play?
12. Examples of receptor cells and effector cells.
13. The properties of extracellular regulatory processes: blood coagulation and complement system activation.
14. How are regulatory loops coupled to one another, and what is the result of such coupling?
15. Why are cells universally small and have low internal volume?

The mechanisms described in this chapter can be experimented with using web applications we provide for the reader's convenience: The NF Organized Systems (NFS, negative feedback system) tool, available at https://nfs.sano.science, where the user may model interconnected systems and their interactions, depending on system parameters—such as receptor sensitivity, effector reaction speed, or the time it takes for signals to travel between the receptor and the effector.

[Wach J, Bubak M, Nowakowski P, Roterman I, Konieczny L, Chłopaś K. Negative feedback inhibition—Fundamental biological regulation in cells and organisms. In: Simulations in Medicine—Pre-Clinical and clinical applications. Ed: Irena Roterman-Konieczna, Walter de Gruyter 2015, pp. 31–56]

Biological entities (cells and organisms) are thermodynamically open systems, which means that they require regulation to maintain a steady state of nonequilibrium. In non-sentient systems regulation must be automatic and based on negative feedback loops. This principle is universally supported by scientific evidence. Given the fact that negative feedback is a precondition of a stable nonequilibrium state, all biological structures may be treated as components of negative feedback loops. Such loops can therefore be interpreted as structural units, representative of specific biological functions. Interrupting a feedback loop at any point results in its malfunction or complete loss of function. Figure 4.1 presents a negative feedback loop with receptor and effector units.

**Fig. 4.1** Automatic control of processes occurring in biological systems. Symbolic depiction of a negative feedback loop with inflow of substrates and release of products

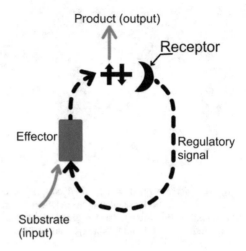

Although the principle of regulation remains unchanged, specific technical solutions (the structure of receptors, effectors, and transmission channels) may differ depending on the circumstances.

## 4.1 The Cell and the Organism

Cells and organisms differ with respect to their regulatory strategies. Clearly, the organism is hierarchically superior to the cell. Its task is to coordinate the function of specialized cells, protect them from harm, and create conditions which promote homeostasis. The relation between a cell and an organism is similar to the citizen-state model. Its hierarchical aspects are quite real and reflected in the structure and function of both entities.

Each living cell is an independent biological unit (with its own source of power), capable of maintaining its internal processes in a steady state through automatic regulation based on negative feedback loops.

Most intracellular signal pathways work by altering the concentrations of selected substances inside the cell. Signals are registered by forming reversible complexes consisting of a ligand (reaction product) and an allosteric receptor complex. When coupled to the ligand, the receptor inhibits the activity of its corresponding effector, which in turn shuts down the production of the controlled substance ensuring the steady state of the system.

Signals coming from outside the cell are usually treated as *commands* (covalent modifications), forcing the cell to adjust its internal processes and enter a new steady state. Although both types of signals (intracellular and extracellular ones) belong to negative feedback loops, they differ with respects to the mechanisms they employ.

Cells are automatic systems, devoid of decision centers. Thus, they are incapable of commanding themselves and can only respond to some external commands. Such commands can arrive in the form of hormones, produced by the organism to

A                              B                              C

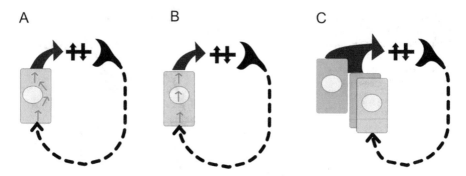

**Fig. 4.2** Three ways in which an effector cell belonging to an organic regulatory pathway may perform its task: (**a**) by regulating the activity of existing proteins (the signal does not pass through the nucleus), (**b**) through gene expression (altering the quantity of proteins—the signal passes through the nucleus), and (**c**) through cell proliferation (both **a** and **b**)

coordinate specialized cell functions in support of general homeostasis (in the organism). These signals act upon cell receptors and are usually amplified before they reach their final destination (the effector). Proper functioning of effector cells in organic regulatory pathways can be ensured in two ways:

1. In a non-dividing cell by:
   A. Activation or inhibition of existing proteins (mostly enzymes) without altering their concentrations
   B. Synthesis or degradation of proteins
2. Through cell proliferation and expression of their biological functions (Fig. 4.2)

Although many regulatory phenomena have not yet been sufficiently studied, current scientific knowledge enables us to formulate some generalizations.

## 4.2  The Principle and Mechanism of Automatic Intracellular Regulation

Basic intracellular signals are expressed through changes in concentrations of reaction substrates and products. This basic form of signaling remains effective so long as the signal does not undergo dilution. By reacting to its own product, a synthesis process may upregulate or downregulate itself, maintaining the genetically programmed concentration of the substance it synthesizes.

The role of negative feedback loop is to stabilize processes whose products are recognized by their own detection subsystems, downregulating the activity of effectors. Automatic control of all internal processes makes the cell an autonomous entity.

## 4.2.1 Cellular Receptors

Each concentration-mediated signal must first be registered by a detector. Once activated, the detector issues another signal, triggering a process which counteracts the observed change. Intracellular detectors are typically based on allosteric proteins.

Allosteric proteins exhibit a special property: they have two stable structural conformations and can shift from one form to the other as a result of changes in ligand concentrations. Examples of such proteins include regulatory enzymes which play a key role in regulating intracellular metabolic pathways. They enable specific binding of reaction products (acting as receptors), but they also exhibit effector-like properties. Most intracellular allosteric enzymes are linked to regulatory functions. Both types of activity (receptor and effector) are interlinked via allosteric effects.

Some regulatory tasks can only be fulfilled by protein complexes due to the fact that the energy cost associated with reconfiguration of subunits is relatively low and can be covered by forming a reversible bond with the ligand. On the other hand, the protein-ligand interaction must result in significant structural rearrangement of the complex, precipitating a significant change in its activity in spite of the relatively low energy associated with such interactions. The allosteric properties of regulatory proteins enable weakly interacting ligands to effectively control biological processes. Their function may be compared to that of converters in electrical circuits, where relatively low currents regulate the action of powerful effector devices.

Allosteric effects are not equivalent to regulatory systems: they merely enable proteins to adapt to changing conditions by undergoing structural rearrangement, resulting in increased or decreased activity; they do not, however, stabilize reactions. This is why allosteric modifications (e.g., in the structure of hemoglobin, which compensates for changes in the availability of oxygen) should be counted among adaptation mechanisms. Allosteric proteins enhance or support regulatory loops but do not replace them. (An exception should be made for certain autonomous proteins which act as receptors as well as effectors, implementing all elements of a regulatory loop—for instance, hexokinase in muscle (see Fig. 4.3). The product of hexokinase—glucose-6-phosphate—is also its inhibitor, responsible for stabilizing the synthesis process.)

Subunits of regulatory enzymes which double as intracellular detectors may be either identical or dissimilar. Subunits which react with the product of a given reaction and perform adjustments by way of allosteric effects are called *regulatory subunits*, while those directly involved in biological functions are known as *functional subunits* (or *catalytic subunits* in the case of enzymes; see Fig. 4.4). If the subunits of an allosteric protein are identical to one another, they may perform receptor and catalytic functions at the same time using different active sites for each function (cumulative feedback inhibition). A classic example is bacterial glutamine synthetase. This enzyme consists of identical subunits (12) which are capable of reacting with various ligands, adjusting the productivity of the entire complex in general. In general the structure of allosteric proteins must enable changes in activity caused by reversible interaction with their own products or with the products of other coordinating processes.

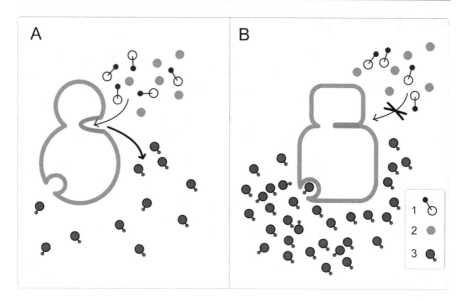

**Fig. 4.3** Symbolic depiction of an allosteric enzyme which implements all elements of a negative feedback loop (hexokinase). A single protein molecule can act as a receptor (recognizing the concentration of glucose-6-phosphate) and an effector (synthesizing additional glucose-6-phosphate molecules). 1, ATP; 2, glucose; 3, glucose-6-phosphate

The concentration of a product (or substrate) which triggers structural realignment in the allosteric protein (such as a regulatory enzyme) depends on the genetically determined affinity of the active site to its ligand. Low affinity results in high target concentration of the controlled substance, while high affinity translates into lower concentration (Fig. 4.5). In other words, high concentration of the product is necessary to trigger a low-affinity receptor (and vice versa).

Most intracellular regulatory mechanisms rely on noncovalent interactions. Covalent bonding is usually associated with extracellular signals, generated by the organism and capable of overriding the cell's own regulatory mechanisms by modifying the sensitivity of receptors (Fig. 4.6). Noncovalent interactions may be compared to *requests*, while covalent signals are treated as *commands*.

Signals which do not originate in the receptor's own feedback loop but modify its affinity are known as steering signals (Fig. 4.7). A controlled regulatory system resembles a servomechanism (i.e., an autonomous, self-regulating mechanical device whose action is subject to external commands). Steering signals play an important role in coordinating biological processes.

Receptor affinity may change as a result of external commands.

According to this definition, intracellular coordinating signals may be described as steering signals; however their interaction with the receptor is noncovalent, unlike the action of signals coming from the organism (the issue of coordination will be discussed in a separate chapter).

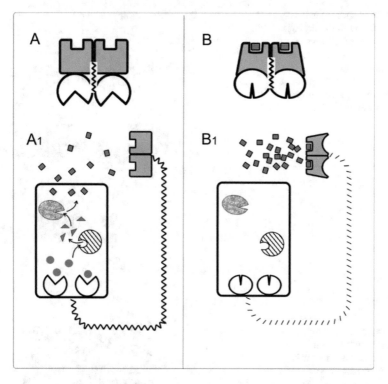

**Fig. 4.4** The action of an intracellular receptor. (**a**) Active regulatory enzyme. (**b**) Regulatory enzyme suppressed by the product. (**a1** and **b1**) Enzyme subunits, shown separately as individual parts of a negative feedback loop. Gray structures, receptor regulatory subunits. White structures, catalytic subunits and enzymes associated with effector activity. The zigzag pattern symbolizes allosteric properties of the protein

Noncovalent interactions—dependent on substance concentrations—impose spatial restrictions on regulatory mechanisms. Any increase in cell volume requires synthesis of additional products in order to maintain stable concentrations. The volume of a spherical cell is given as $V = 4/3 \, \pi * r^3$, where $r$ indicates cell radius. Clearly, even a slight increase in $r$ translates into a significant increase in cell volume, diluting any products dispersed in the cytoplasm. This implies that cells cannot expand without incurring great energy costs. It should also be noted that cell expansion reduces the efficiency of intracellular regulatory mechanisms because signals and substrates need to be transported over longer distances. Thus, cells are universally small, regardless of whether they make up a mouse or an elephant.

## 4.2.2 Cellular Effectors

An effector is an element of a regulatory loop which counteracts changes in the regulated quantity.

**Fig. 4.5** Negative feedback loop receptors with varying degrees of affinity to their product: high (**a**), moderate (**b**), and low (**c**) concentrations of the product resulting from variations in receptor affinity

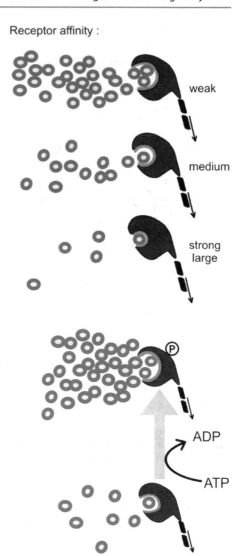

Receptor affinity :

weak

medium

strong
large

**Fig. 4.6** Schematic view of the changes in receptor affinity resulting from covalent modification of its structure (phosphorylation)

ADP

ATP

   Cellular effectors usually assume the form of degradation processes or feedback-controlled synthesis mechanisms. In both cases the concentration of the regulated product is subject to automatic stabilization and control. From the point of view of the entire organism, each cell can be treated as a separate effector.

   Synthesis and degradation of biological compounds often involve numerous enzymes acting in sequence. The product of one enzyme is a substrate for another enzyme. With the exception of the initial enzyme, each step of this cascade is controlled by the availability of its substrate and does not require separate allosteric regulators (Fig. 4.8).

**Fig. 4.7** Schematic view of a negative feedback loop coupled to a control unit (model of servomechanism)

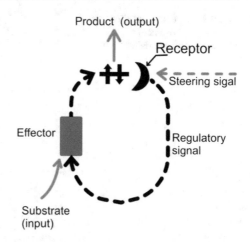

The effector consists of a chain of enzymes, each of which depends on the activity of the initial regulatory enzyme (aspartate transcarbamoylase—ATCase) as well as on the activity of its immediate predecessor which supplies it with substrates.

The function of all enzymes in the effector chain is indirectly dependent on the initial enzyme, allosterically linked to a receptor subsystem. This coupling between the receptor and the first link in the effector chain is a universal phenomenon. It can therefore be said that the initial enzyme in the effector chain is, in fact, a regulatory enzyme. (Note that by "initial enzyme" we mean the enzyme whose product unambiguously triggers a specific chain of reactions.)

The product of the effector chain (or its functional derivatives) forms reversible complexes with the regulatory enzyme, controlling the production process, and through it—the function of the effector itself. By binding its assigned product, the regulatory enzyme acts as an intracellular receptor in addition to catalyzing the reaction.

Most cell functions depend on enzymatic activity. Even nonenzymatic processes (e.g., the action of motor proteins) are indirectly mediated by enzymes. It seems that a set of enzymes associated with a specific process which involves a negative feedback loop is the most typical form of an intracellular regulatory effector. Such effectors can be controlled through activation or inhibition of their associated enzymes.

A typical eukaryotic cell is, however, largely incapable of affecting enzymes other than by synthesizing their activators or inhibitors. Due to the fact that organisms actively strive to maintain internal homeostasis, enzymatic control cannot be enforced through major changes in temperature, pH, or substance concentrations. Instead, control rests upon the interaction between allosteric enzymes and their receptors. The dynamics of metabolic pathways depend on the concentrations of their own products and substrates (regulation) as well as on the products of other reactions for coordination (steering).

Certain processes involve a somewhat different enzymatic effector control mechanism which exploits changes in the concentration of active enzymes through

**Fig. 4.8** Regulation of CTP
(cytidine triphosphate)
synthesis in a bacterial cell

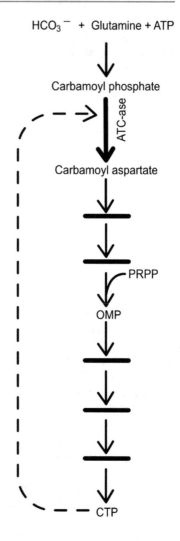

regulated synthesis of proteins (gene expression). A negative feedback loop which alters the concentrations of proteins (or their products) in order to control effectors was first described by F. Jacob and J. Monod, who studied the so-called lactose operon in *E. coli* bacteria (Fig. 4.9). This operon includes a receptor called a *repressor* (or, more specifically, its detection subunit). The repressor binds its assigned product (or substrate) and undergoes allosteric realignment which causes it to gain (or lose) the ability to attach to the operator subunit, thereby activating (or inhibiting) the function of polymerase. In this way the effector may be inactivated once the concentration of the product exceeds a genetically determined level, thus preventing excessive accumulation of products. Similar modulation of transcription is also observed in eukaryotic organisms, although in this case it must occur simultaneously in many areas of the genome, as the encoded information concerning

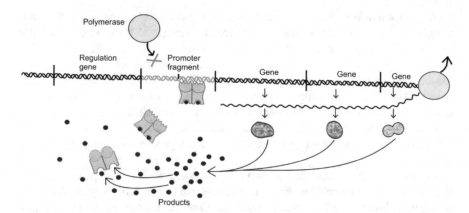

**Fig. 4.9** The bacterial operon—a negative (self-regulated) feedback loop with transcription and translation as its basic effector mechanisms

the given activity is often partitioned into different chromosomes. Instead of repressors, eukaryotes employ so-called transcription factors—having special structural matrices (zinc fingers or leucine zippers) capable of attaching to DNA strands in response to external signals.

Receptor activity is exhibited by repressor subunits which react with the product (called an inductor) and allosterically modify the repressor's ability to bind to DNA.

Transcription activation processes have been extensively studied; however most of these studies involve extracellular signals, i.e., signals generated by the organism. Transduction of such signals is ensured by membrane receptors and other specific transduction factors (such as G proteins). It is not clear, however, how the cell regulates transcription for its own purposes. The simplest model seems to be an operon-like construct where the protein product (e.g., enzyme) directly interacts with the transcription factor, upregulating or downregulating transcription as required. Nevertheless, our knowledge of this mechanism is not yet sufficient to attempt generalizations.

Another related model involves partial or complete appropriation of the own intracellular machinery for interpreting hormonal signals sent by the organism. This is realized by auto- or paracrine signaling (see Figs. 4.14, 4.15, 4.16, and 4.18).

## 4.3   Regulatory Coupling Between Cells and Organisms: Hierarchical Properties of Regulation

Automatic regulation enables cells to act as autonomous units, which means that—in given suitable environmental conditions—cells may survive on their own. In an organism, however, the autonomy of individual cells is severely restricted. The relation between the cell and the organism is hierarchical, and although the organism only controls specific areas of the cell's activity, its demands cannot be met without the aid of basic intracellular processes.

A question arises: how can an involuntary automatic system (i.e., the cell) integrate with the organism's own regulatory mechanisms? This issue can be better explained by comparing the cell to a simple automaton, namely, a refrigerator. The refrigerator includes an automatically controlled thermostat, based on a negative feedback loop with a temperature sensor coupled to a heat pump. Like most automatons, it is an autonomous system, capable of stabilizing its own internal temperature. However, most refrigerators also permit external control by an operator who may change the desired temperature. Operator intervention occurs by way of interacting with a separate panel (receptor) which registers and interprets steering signals.

Assuming that the stabilization curve for negative feedback loops is sinusoid-shaped, the effect of control is similar to that depicted in Fig. 4.10. A good biological example is fever, where bacterial toxins act upon organic temperature receptors which in turn modulate body temperature. Once the bacteria are cleared from the organism, the temperature stabilization loop reverts to its original parameters. Such changes in receptor sensitivity are perceived as a cold or heat sensation which persists until the desired temperature is reached.

While refrigerator settings can be modified at will, signals issued by the organism are themselves products of other hierarchically superior regulatory loops which control the functions of the organism as a whole. This mechanism is depicted in Fig. 4.11. Hormones which act upon cells are, by their nature, steering signals: they enforce the superiority of the organism in relation to individual cells. The stabilization loop of a homeostatic parameter becomes a source of control if its signals are capable of affecting cellular receptors by altering their structure or introducing other permanent changes (such as the formation of complexes). This effect can be attained, e.g., by covalent modification of receptor units.

This schematic relation is an expression of the cell-organism hierarchy.

A well-studied example of the presented mechanism involves regulation of blood glucose levels through degradation and synthesis of glycogen in hepatic cells. In this case an intracellular regulatory loop maintains the desired low concentration of glucose inside the cell (as indicated by glucose-6-phosphate). Glucose can be

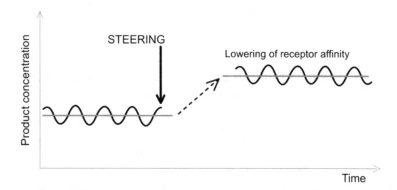

**Fig. 4.10** Changes in stabilization thresholds caused by steering signals

**Fig. 4.11** Changes in cellular receptor sensitivity (in this case—downregulation, resulting in increased tolerance to the product) caused by a hormonal steering signal which originates outside of the cell's own regulatory loops

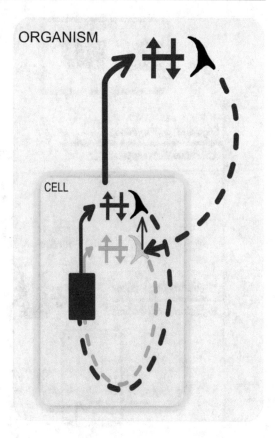

obtained from glycogen in the course of phosphorolysis, which is dependent on the concentration of ATP. Signals generated by a hierarchically superior control loop may force the cell to intensify its metabolism by altering the sensitivity of an allosteric enzyme—glycogen phosphorylase. Covalent (hormonally induced phosphorylation) modifications to this enzyme increase its tolerance to the product it controls (see Fig. 4.12).

Hormonal modification significantly reduces the affinity of the receptor to glucose-6-phosphate. As a result, the concentration of glucose-6-phosphate (and, consequently, of glucose) increases. Excess glucose is expelled into the bloodstream where it can compensate for the deficiency which triggered the initial hormonal steering signal.

In summary, regulatory loops of the organism rely on the effector activity of individual cells through (1) modifying the activity of specific intracellular proteins, (2) degradation or synthesis of proteins connected with the given activity, and (3) cell proliferation or programmed cell death. Hormone-induced activity, although specific, affects also the entire biological machinery of the cell.

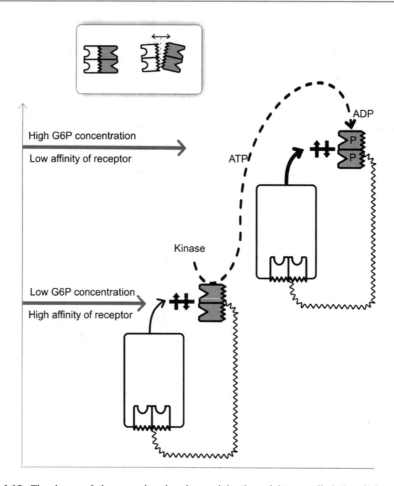

**Fig. 4.12** The change of glycogen phosphorylase activity through hormonally induced phosphorylation. Changes in sensitivity of the receptor result in adjustment of glucose-6-phosphate concentrations, depicted on the vertical axis. Gray structures represent receptor units

## 4.4   Regulatory Mechanisms on the Organism Level

The organism is a self-contained unit represented by automatic regulatory loops which ensure homeostasis. Its program is expressed in the structure of its receptors. Effector functions are conducted by cells which are usually grouped and organized into tissues and organs. Signal transmission occurs by way of body fluids, hormones, or nerve connections. Cells can be treated as automatic and potentially autonomous elements of regulatory loops; however their specific action is dependent on the commands issued by the organism. This coercive property of organic signals is an integral requirement of coordination, allowing the organism to maintain internal homeostasis.

Coercive action can be achieved through (1) significant amplification of signals issued by the organism to its effector cells, (2) covalent nature of modifications triggered by organic signals, and (3) a possible slow degradation of organic signals which have reached the cell. The effects of these mechanisms can be compared to an army drill, where the instructor enforces obedience by persistently shouting orders.

Activities of the organism are themselves regulated by their own negative feedback loops. Such regulation differs however from the mechanisms observed in individual cells due to its place in the overall hierarchy and differences in signal properties, including in particular:

1. Significantly longer travel distances (compared to intracellular signals)
2. The need to maintain hierarchical superiority of the organism
3. The relative autonomy of effector cells

In order to remain effective, the signal coming from the organism must be amplified and encoded; moreover its activity should be independent of its concentration (through specific covalent bonding).

It is also necessary to provide signal deactivation mechanisms which can shield effectors from undue stress caused by the outliving hierarchical signal (note that the effectors in organic regulatory pathways are living cells which require protection).

Most organic signals travel with body fluids; however if a signal has to reach its destination very rapidly (for instance, in muscle control), it is sent via the nervous system and becomes only humoral at entering the cell.

As mentioned above, organic signals act upon cellular regulatory loops which control cell development and/or homeostasis. Responding to such signals and fulfilling the requested tasks require the cell to adjust its internal metabolism.

## 4.4.1  Signal Encoding

The relatively long distance traveled by organic signals (compared to intracellular ones) calls for amplification. As a consequence, any errors or random distortions in the original signal may be drastically exacerbated.

A solution to this problem comes in the form of encoding, which provides the signal with sufficient specificity while enabling it to be selectively amplified. This situation can be compared to talking over a shortwave radio. When engaged in a normal conversation, we are usually facing our interlocutor who can hear us clearly, despite any ambient noise. In order to communicate over longer distances, we need to use a radio transceiver which encodes our voice as an electromagnetic wave (unlike acoustic waves, electromagnetic signals are unaffected by ambient noise). Once received and amplified by the receiver, the signal may be heard by the other party. Note that a loudspeaker can also assist in acoustic communication, but due to

**Fig. 4.13** Encoding and decoding of organic signals in a negative feedback loop (regulation of blood glucose levels)

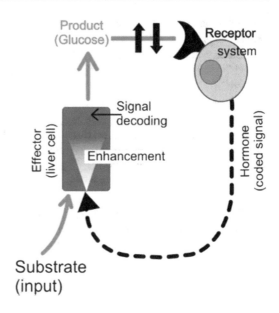

the lack of signal encoding, it cannot compete with radios in terms of communication distance.

The same reasoning applies to organism-originated signals, which is why information regarding blood glucose levels is not conveyed directly by glucose but instead by adrenalin, glucagon, or insulin.

Information encoding is handled by receptors and hormone-producing cells. Target cells are capable of decoding such signals, thus completing the regulatory loop (Fig. 4.13).

The receptor system is associated with the endocrine gland cell.

Hormonal signals may be effectively amplified because the hormone itself does not directly participate in the reaction it controls—rather, it serves as an information carrier. If blood glucose concentration was to manifest itself purely through changes in glucose levels (without hormonal action), local distortions and random effects would—following intracellular amplification—interfere with correct interpretation of the input signal. Thus, strong amplification invariably requires encoding in order to render the signal sufficiently specific and unambiguous.

The above mechanism is ubiquitous in nature. An interesting example is the *E. coli* lactose operon, exploiting as an inducer allolactose which is attuned as a tiny component to lactose instead of the somewhat more expected lactose itself (the proper substrate). Registering an increase in the concentration of allolactose serves as direct evidence that lactose levels have also increased. As a consequence, the operon activates gene expression for additional enzymes, adjusting the cell's energy management mechanisms.

Allolactose has the properties of an encoded signal. If the receptor were to react directly to lactose, it would have to cope with the relative abundance of this substance by lowering its own sensitivity—this, however, would also decrease its specificity and increase the probability of errors.

## 4.4.2 Signal Amplification

As mentioned above, one of the key differences between intracellular and organism-originated signals is the distance each has to travel. Long-distance communication introduces the need for signal amplification.

An important subgroup of regulatory components involved in organism-originated homeostasis processes is represented by amplifiers. Unlike organisms, cells usually do not require amplification in their internal regulatory loops—even the somewhat rare instances of intracellular amplification only increase signal levels by a small amount.

Without the aid of an amplifier, messengers coming from the organism level would need to be highly concentrated at their source, which would result in decreased efficiency as all such molecules need to be synthesized first (and then degraded once the signal has served its purpose). The need for significant amplification is also tied to the specific properties of organism-originated signals which must override the cell's own regulatory mechanisms.

Two types of amplifiers are observed in biological systems:

1. Cascade amplifier
2. Positive feedback loop

## 4.4.3 Cascade Amplifier

A cascade amplifier is usually a collection of enzymes which perform their action in strict sequence. This mechanism resembles multistage (sequential) synthesis or degradation processes; however instead of exchanging reaction products, amplifier enzymes communicate by sharing activators or by directly activating one another. Cascade amplifiers are usually contained within cells. They often consist of kinases. A classic example is the adenylate cyclase cascade in the glucagon synthesis pathway (response to decreased blood glucose levels), whose individual stages are invoked through phosphorylation (Fig. 4.14). Activation of amplifier stages may also occur by way of protease or esterase interaction, release of calcium ions, etc. (Fig. 4.15). Figure 4.16 presents the amplification of growth factors. Amplification is also involved in gene expression as each stage amplifies information via repeated transcription and interpretation of genetic code as well as through prolonged reuse of the resulting proteins to synthesize additional products or perform some other activity (Fig. 4.17).

The effector cell regulates the amplifier by synthesizing its substrates (such as ATP). Amplification effects occurring at each stage of the cascade contribute to its final result. In the case of adenyl cyclase, the amplified hormone allows for synthesis

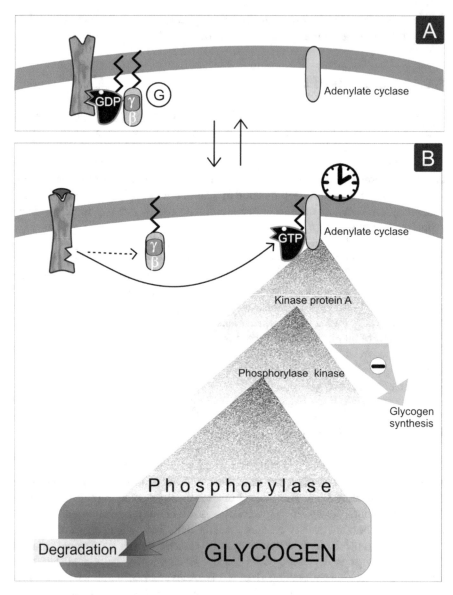

**Fig. 4.14** Amplification mechanisms associated with intracellular hormonal action: adenylate cyclase cascade. Gray triangles represent staged amplification

of approximately $10^4$ cyclic AMP molecules before it is inactivated. While the kinase amplification factor is estimated to be on the order of $10^3$, the phosphorylase cascade results in $10^{10}$-fold amplification. It is a stunning value, though it should also be noted that the hormones involved in this cascade produce particularly powerful effects.

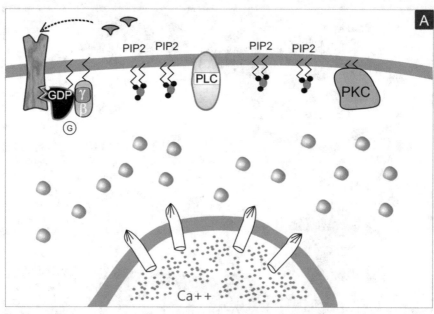

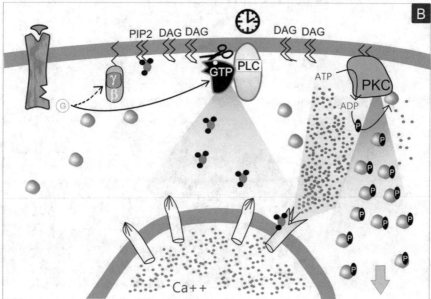

**Fig. 4.15** Amplification mechanisms associated with intracellular hormonal action: phospholipase C cascade. (**a**) Conditions prior to hormonal activation of the cell. (**b**) The cell in its activated state. Gray triangles represent amplification. DAG, 1,2-diacylglycerol; PIP$_2$, phosphatidyl-inositol-4,5-biphosphate; IP$_3$, inositol-1,4,5-triphosphate; PKC, protein C kinases; G, proteins; PLC, phospholipase C

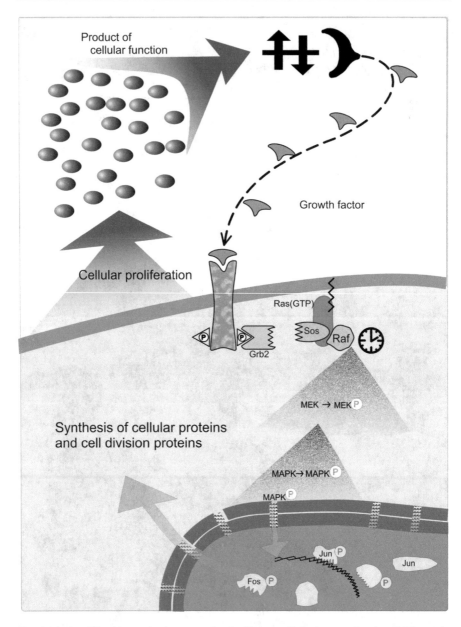

**Fig. 4.16** Amplification mechanisms associated with intracellular hormonal action: RAS protein cascade activated by the growth factor

Due to their low volume, cells usually have no need for signal amplification. Even so, amplification may play a certain role in special cases, such as autocrine signaling, where a signal produced by the cell returns to it and is recognized by a receptor. This

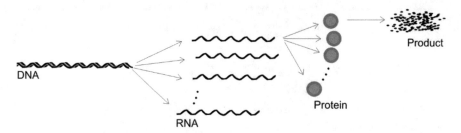

**Fig. 4.17** Amplification in gene expression

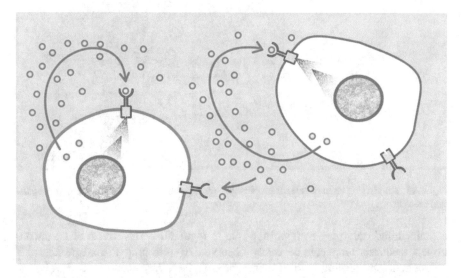

**Fig. 4.18** Autocrine signaling—repurposing the hormonal signal pathways for amplification of the cell's own signals

situation corresponds to a self-issued command. Autocrine amplification becomes necessary in transcription-inducing processes as it enables the cell to "repurpose" mechanisms normally used to amplify external signals (Fig. 4.18).

## 4.4.4   Positive Feedback Loop

A positive feedback loop is somewhat analogous to a negative feedback loop; however in this case the input and output signals work in the same direction—the receptor upregulates the process instead of inhibiting it. Such upregulation persists until the available resources are exhausted.

Positive feedback loops can only work in the presence of a control mechanism which prevents them from spiraling out of control. They cannot be considered self-

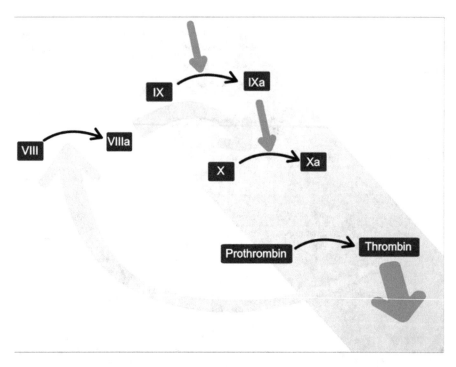

**Fig. 4.19** Amplifying an extracellular process through a positive feedback loop—blood coagulation cascade

contained and only play a supportive role in regulation. The effects of an uncontrolled feedback loop can be easily observed by attaching a loudspeaker to a microphone and placing both of them close together. Any sound registered by the microphone is amplified and transmitted over the loudspeaker, further increasing the input volume and causing a runaway feedback reaction which manifests itself as the familiar screeching noise.

In biological systems positive feedback loops are sometimes encountered in extracellular regulatory processes where there is a need to activate slowly migrating components and greatly amplify their action in a short amount of time. Examples include blood coagulation and complement factor activation (Figs. 4.19 and 4.20).

Positive feedback loops may play a role in binary control systems which follow the "all or nothing" principle. Due to its auto-catalytic character, the positive feedback loop is used for construction of binary switching mechanisms and governing cell fate decisions.

Positive feedback loops are often coupled to negative loop-based control mechanisms. Such interplay of loops may impart the signal with desirable properties, for instance, by transforming a flat signal into a sharp spike required to overcome the activation threshold for the next stage in a signaling cascade. An example is the ejection of calcium ions from the endoplasmic reticulum in the phospholipase C

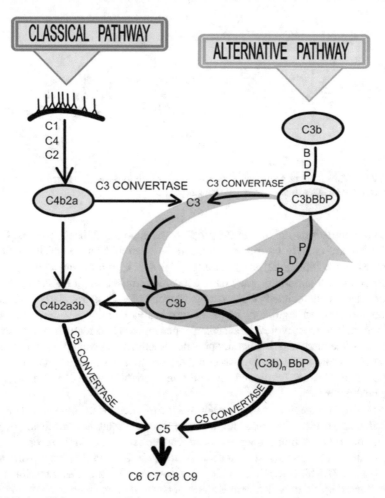

**Fig. 4.20** Amplifying an extracellular process through a positive feedback loop—complement factor activation cascade

cascade, itself subject to a negative feedback loop. Positive feedback is also observed in the activation of cyclins.

Feedback-based amplification of weak signals may also double as intracellular memory. It is believed that the high rate of activation and deactivation of signals in nerve cells relies on positive feedback mechanisms.

## 4.4.5   Signal Attenuation

Strong signal amplification carries an important drawback: it tends to "overshoot" its target activity level, causing wild fluctuations in the process it controls. While

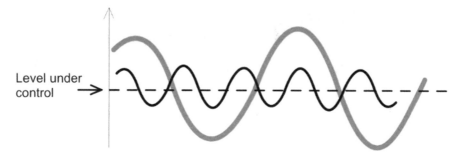

**Fig. 4.21** Stabilization of a substance concentration or activity level as a function of a negative feedback loop: changes in amplitude and frequency

sinusoid fluctuations are a natural property of many stabilized processes, high amplitudes are usually considered undesirable even if average values remain in agreement with biological programming (Fig. 4.21). Significant deviations from the norm, though temporary, may have a negative impact on biological systems.

The danger associated with strong fluctuations is particularly evident in organism-originated regulatory loops. Nature has evolved several means of signal attenuation. The most typical mechanism superimposes two regulatory loops which affect the same parameter but act in opposite directions. An example is the stabilization of blood glucose levels by two contradictory hormones: glucagon and insulin. Similar strategies are exploited in body temperature control and many other biological processes.

Each cellular receptor acts upon a single type of effector: its input signal can only work in one specific direction, for instance, by activating the synthesis of some product or counteracting a decrease in its concentration. A single receptor cannot issue contradictory signals, simultaneously upregulating and downregulating a process (Fig. 4.22). This limitation usually does not hamper intracellular machinery due to the limited space in which reaction components are contained and the efficiency with which they can be processed. However, significant fluctuations may be observed in certain slow-acting processes such as gene expression where the persistence of mRNA may sometimes induce overexpression. Maintaining the programmed level of activity, free of undue fluctuations, requires proper degradation and removal of intermediate products.

The attenuation mechanism involved in the action of the bacterial operon has been well studied. It is able to recognize the concentration of the product prior to transcription necessary for its synthesis and prevent its excessive buildup as a result of recycling transcripts.

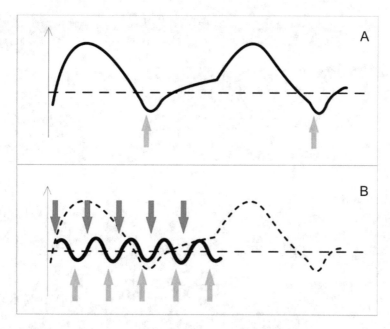

**Fig. 4.22** Fluctuations in a controlled quantity, given (**a**) unidirectional detection and control and (**b**) bidirectional control based on two contradictory negative feedback loops

## 4.4.6  Signal Inactivation

The coercive properties of signals coming from the organism carry risks associated with the possibility of overloading cells. The regulatory loop of an autonomous cell must therefore include an "off switch," controlled by the cell.

An autonomous cell may protect itself against excessive involvement in processes triggered by external signals (which usually incur significant energy expenses). Once acknowledged, hormones must be quickly inactivated.

It is natural to expect that the inactivation mechanism will be found along the signal pathway (before the position of amplifier). The action of such mechanisms is usually timer-based, meaning that they inactivate signals following a set amount of time. A classic example of an inactivator is found in protein G (see Figs. 4.14 and 4.15). This protein is activated by hormonal signals, but immediately upon activation it also triggers the corresponding inactivation process. Protein G activation occurs by swapping GDP for GTP and by subsequent reorganization of its subunits. Inactivation proceeds in the opposite direction through autocatalysis (the protein exhibits GTPase activity). The time it takes to complete the hydrolysis process automatically determines the duration of the signal. A similar property is observed in Ras proteins, where GTPase activity of the initiation complex eventually interrupts the signal (Fig. 4.23). We should also note the association between receptor phosphorylation and phosphatase activity in many other processes.

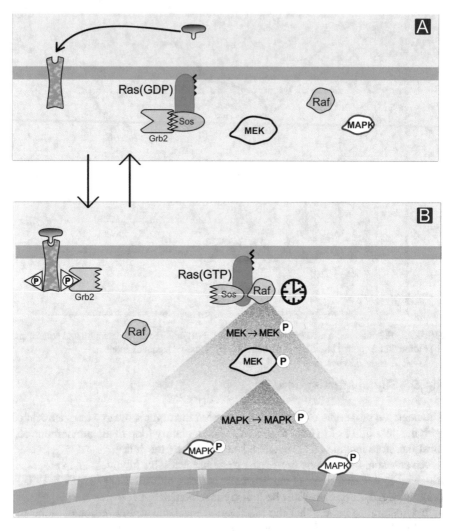

**Fig. 4.23** Signal inactivation (RAS protein). (**a**) Conditions prior to activation and (**b**) hormone-activated complex (the clock symbolizes the "off timer" which depends on GTPase activity). Signal inactivation by protein G is depicted in Figs. 4.14 and 4.15. RAS, RAS proteins; Grb2 and Sos, proteins assisting RAS; Raf, MEK, and MAPK, amplifier kinases

Although all proteins which undergo phosphorylation eventually revert to their initial state via phosphatase activity, the phosphorylation/dephosphorylation effects are particularly pronounced at the entry point of the amplifying cascade which should therefore be considered the proper inactivation place. Signal inactivation may also occur as a result of receptor pinocytosis (Fig. 4.24). The ability to interrupt signals protects cells from exhaustion. Uncontrolled hormone-induced activity may have detrimental effects upon the organism as a whole. This is observed, e.g., in the

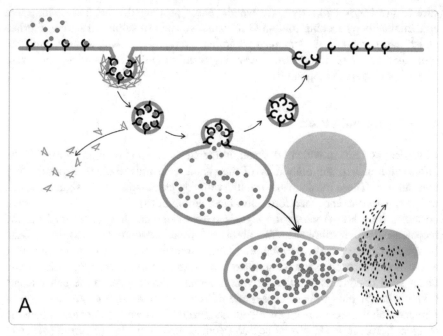

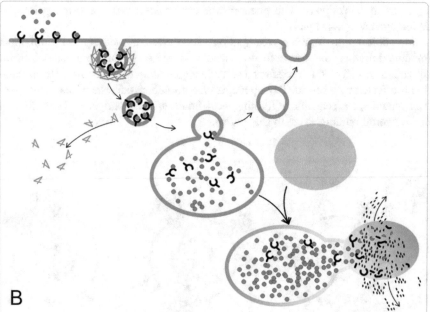

**Fig. 4.24** Surface receptor density controlled by degradation and synthesis. (**a**) Degradation of ligands and (**b**) degradation of ligands and receptors

case of the *Vibrio cholerae* toxin which causes prolonged activation of intestinal epithelial cells by locking protein G in its active state (resulting in severe diarrhea which can dehydrate the organism). Similar dysregulation is associated with the prolonged action of acetylcholine resulting from inhibition of acetylcholinesterase by phosphoorganic compounds.

### 4.4.7   Discrimination

Biological systems in which information transfer is affected by high entropy of the information source and ambiguity of the signal itself must include discriminatory mechanisms. These mechanisms usually work by eliminating weak signals (which are less specific and therefore introduce ambiguities). They create additional obstacles (thresholds) which the signals must overcome. A good example is the mechanism which eliminates the ability of weak, random antigens to activate lymphatic cells. It works by inhibiting blastic transformation of lymphocytes until a so-called receptor cap has accumulated on the surface of the cell (Fig. 4.25). Only under such conditions can the activation signal ultimately reach the cell nucleus (likely with the aid of phosphatase) and initiate gene transcription. Aggregation of immunoglobulin receptors on the membrane surface is a result of interaction with the antigen; however weak, reversible nonspecific interactions do not permit sufficient aggregation to take place. This phenomenon can be described as a form of discrimination against weak signals.

In addition to antigen recognition and effector activity of individual immunoglobulins, another activation threshold exists in the complement factor activation cascade, which requires a suitable aggregation of antibodies. Discrimination is effected by a special T lymphocyte receptor which consists of many independent subunits, each capable of forming weak bonds with the antigen and contributing to the overall binding strength (Fig. 4.26).

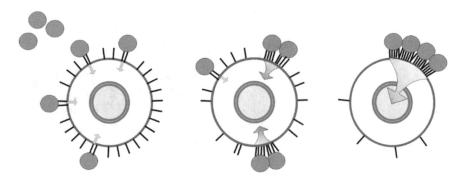

**Fig. 4.25** Hypothetical discrimination mechanism preventing the randomly induced blastic transformation in B cells. Accumulation of receptors is required for signal transduction to take place. Gray circles represent antigens, while lines represent surface receptors

**Fig. 4.26** Schematic structure of the T lymphocyte receptor consisting of many weakly binding molecules which together provide sufficient antigen binding strength. This structure provides discrimination against random, nonspecific signals. CD28, LFA-1, CD4, CD3, TCR, CD2, B7, ICAM-1, MHC II, and LFA-3—cell receptors and markers

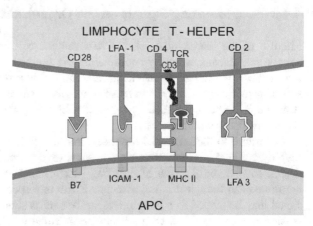

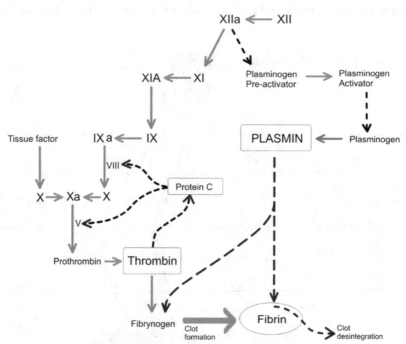

**Fig. 4.27** Fibrinolysis as a discriminator against excessive clotting beyond the specific site of damage

Discrimination may also be linked to effector activity. One example is the restricted propagation of blood clots beyond the specific site of vessel damage. Blood coagulation automatically triggers fibrinolysis, which counteracts further clotting by degrading fibrin molecules (Fig. 4.27).

### 4.4.8   Coordinating Signals on the Organism Level

Mindful of the hierarchical relation between the cell and the organism, we can assume that hormones act by coordinating the function of cells in order to ensure homeostasis. Hormonal signals supersede intracellular regulatory mechanisms and subordinate the needs of cells to those of the organism as a whole. This hierarchical structure is further reinforced by signal amplification and covalent modification of signal mediators within the cell.

According to the presented criteria, hormones follow a specific plan of action, roughly outlined in Fig. 4.28, which presents several transduction pathways. It also depicts some atypical means of signaling—for instance, nitric oxide, which acts as a hormone even though it is not associated with any specific membrane receptor. A signal may be treated as a hormone regardless of its source, as long as it originates outside the cell and affects intracellular machinery. Hormonal control may be effected over long distances (via the bloodstream) or short distances, when the signal

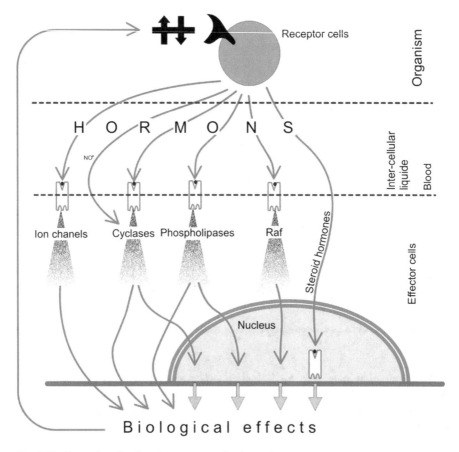

**Fig. 4.28**  Examples of various hormone transduction pathways

comes from adjacent cells (e.g., growth factors or morphogens). Nitric oxide permeates the cellular membrane and activates the guanylyl cyclase cascade. The specificity of this unusual hormone is probably a result of the proximity of its source to the effector mechanism as well as its rapid rate of its degradation.

Another atypical means of signaling is observed in steroids, whose receptors are located inside the cell and—once activated—can perform the role of transcription modulators. The coercive nature of steroids is supported by their persistent action inside the cell (compared to hormones which only interact with surface receptors) as well as their high affinity to receptors (equilibrium constant in the $10^9$ range). It should also be noted that biological processes induced by steroids are usually slow-paced and therefore do not require strong amplification. Hormonal signaling is itself a relatively slow process—signals which need to reach their destination very rapidly (for instance, signals controlling muscle contractions) are usually transmitted via nerve connections.

As mentioned above, if the organism is to coordinate the action of various tissues, its signals must override the autonomous regulatory mechanisms of individual cells.

## 4.4.9 Extracellular Process Control

Most enzymes reside inside cells, acting as effectors in intracellular pathways. There are, however, processes which require cells to release enzymes into the extracellular space. These include food metabolism, blood coagulation, complement factor activation, and some tissue-related mechanisms. Under such circumstances the cell effectively relinquishes control of the given process. While enzymatic activity can be easily regulated inside the cell (for instance, by automatically controlling the concentration of its product), external processes admit no such regulation and can potentially prove dangerous (this applies to, e.g., proteolytic enzymes). Occasionally, enzymes may enter the outer cell space as a result of cell degradation or diffusion from the gastrointestinal tract.

Inadvertent (but manageable) enzyme "leaks" are dealt with by specialized natural inhibitors called serpins which exist in the bloodstream specifically for that purpose. Proper regulation only applies to enzymes secreted by cells in programmed biological processes. As a general rule, such enzymes are released in their inactive state. They can be activated by a specific signal once the organism has taken steps to protect itself from uncontrolled proliferation of active forms.

Activation of gastrointestinal enzymes is therefore coupled to processes which protect the intestinal wall from being attacked. In blood coagulation, enzymatic activity is tightly linked to the developing clot, while in the complement factor activation cascade, degradation processes are limited to antigen-antibody complexes present on the surface of the antigen cell. The principle of production and secretion of inactive enzymes is evident in extracellular mechanisms where controlling such enzymes may be difficult; however it can also be observed in certain intracellular pathways where some enzymes are activated ab initio to fulfill specific tasks. This

situation occurs, e.g., in kinases (activated through phosphorylation) or in proteolytic enzymes (especially caspases) tightly connected with apoptosis.

## 4.4.10 Cell Population Control

Basic cell population control mechanisms include proliferation by division (mitosis)
and programmed cell death (apoptosis). Both processes must be controlled by
extracellular signals because they produce important effects for the organism as a
whole.

Cell division is governed by the so-called growth factors. Once triggered, the
division process may proceed autonomously, progressing through successive
checkpoints of the mitosis program (Fig. 4.29).

If the process consists of several stages, control must be sequential, i.e., each
stage starts with a checkpoint which tests whether the previous stage has completed
successfully.

Cell division requires the entire genome to be copied, which involves not just
DNA replication but also synthesis of additional DNA-binding proteins. This
so-called S phase comes at a tremendous cost to the cell's resources. It is followed
by the $G_2$ phase, which prepares the cell for actual division (itself occurring in the M
phase). All processes associated with division are subject to strict control. Each stage
must fully complete before the next stage is triggered.

Phosphorylation and dephosphorylation (mediated by kinases and phosphatases)
appear to effect control over the mitosis process. However, each phase of mitosis is

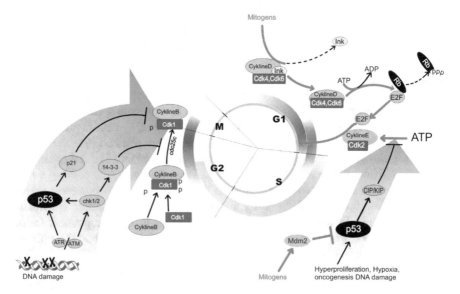

**Fig. 4.29** Schematic depiction of cell division (mitosis). Colored areas indicate the activation
pathway. Gray belts show the activity of cyclins (A to E)

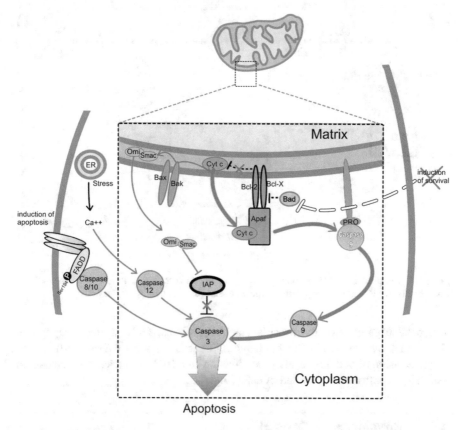

**Fig. 4.30** Schematic depiction of apoptosis. Activation pathways marked in color

ushered in by special proteins called cyclins. These are synthesized in parallel with other division-related structures and broken down shortly thereafter.

The cell may maintain a steady state ($G_0$ phase) for a long time. Entering the $G_1$ phase signals readiness for division; however the transition from $G_0$ to $G_1$ is reversible. Cells protect themselves against accidental division by raising the activation threshold. This mechanism is mediated by a protein known as p53, which can also inhibit division in later phases, should irregularities occur.

Cell division is counterbalanced by programmed cell death. The most typical example of this process is apoptosis (Fig. 4.30). Apoptosis occurs as a result of proapoptotic signals (such as the *tumor necrosis factor-alpha* (TNF-α)) which pierce the mitochondrial membrane and release the contents of mitochondria into the cytoplasm. Once released, mitochondrial cytochrome C activates specific apoptotic enzymes called caspases, which proceed to degrade cell organelles and effectively kill the cell.

Each cell is prepared to undergo controlled death if required by the organism; however apoptosis is subject to tight control. Cells protect themselves against

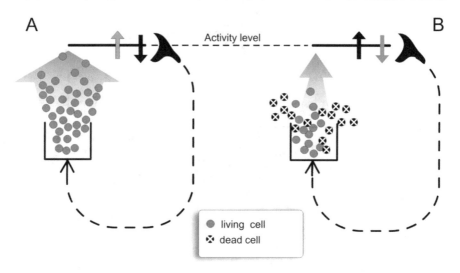

**Fig. 4.31** Simplified view of cell population management via controlled proliferation and destruction of cells. (**a**) Reacting to population deficiency and (**b**) eliminating surplus cells

accidental triggering of the process via IAP proteins. Only strong proapoptotic signals may overcome this threshold and initiate cellular suicide (Fig. 4.30).

Supervision of both processes—cell division and death—enables the organism to maintain a controlled population of cells (Fig. 4.31).

## 4.5    Development Control

Biological regulation relies on the assumption that processes can be automatically controlled via negative feedback loops. Such regulation appears useful both for the organism and for individual, mature cells which strive to maintain a steady state. However, it cannot be readily applied to development control where the goal is to increase the population of cells and cause irreversible changes in their structure. Developmental changes are controlled at specific checkpoints during the process and following its conclusion (see Chap. 3, Fig. 3.13). Signals which guide development are themselves an expression of an evolving genetic blueprint. They control cell differentiation as well as the shaping of organs and the entire organism.

Development control algorithms are implemented through sequential induction of cell proliferation and differentiation, where cells and signals are subject to hierarchical management. Information may be conveyed through direct interactions between adjacent cells or by hormonal signals (typically acting at short ranges). A simple model which may serve to explain this process is the formation of a photoreceptor in the insect eye ommatidium. This element normally consists of eight separate cells, controlled by a specialized cell called R8. The R8 cell is also sometimes called *BOSS*—an acronym of the term *Bride Of SevenlesS*, referring to a specific mutation which interferes with the differentiation of the seventh cell and reinforces the supervisory role of the eighth cell. Figure 4.32 presents the differentiation algorithm for the development of the ommatidium photoreceptor, showing the differences in processes occurring on each side of the R8 cell. This differentiation mechanism results in asymmetric arrangement of the photoreceptor element, which in turn enables the insect to determine which direction the light is coming from. The staged, programmed maturation of photoreceptor cells is an example of a generalized process from which specialized organs and structures may emerge.

## 4.6 Basic Principles of Regulation in Biology

In general, it can be said that any activity which ensures the stability of biological processes may be called regulation and that regulation is dependent on negative feedback loops. Regulatory mechanisms work to maintain cellular and organic homeostasis, utilizing receptor systems which determine, e.g., the concentrations of various substances in blood or in cell cytoplasm.

On the other hand, signals which alter the default stabilization levels are called steering signals. They originate beyond the loop which they control and may be divided into intracellular signals (allosteric effectors) and hormonal signals which guide the specific activity of the cell. Steering signals may also supervise the development process by triggering the formation of new structures and expression of their functions.

From the point of view of kinetics, biological control mechanisms may be divided into three groups:

1. Response control, characteristic of metabolic processes where the intensity of a given process closely follows the concentration of a regulatory hormone. Examples include the reaction of hepatic cells to insulin and glucagon. Such processes can be roughly compared to the action of servomechanisms in a power steering system.
2. Extremal control, where the signal releases or inhibits a specific process, e.g., secretion of stomach acid, cell division, apoptosis, sexual maturation, etc. An appropriate mechanical counterpart is the activation/deactivation of a light switch by a photodetector.
3. Sequential control. This type of control is observed in development and growth. Steering signals are issued sequentially, according to a specific algorithm. Each stage of development is triggered once the previous stage has concluded and a checkpoint has been reached. In the developing embryo, this involves activation of successive gene packets, while in cell development, it applies to the action of cyclins. Sequential control can also be observed in many macroscopic devices, for instance, in washing machines and assembly-line production environments. In biological systems it is usually effected by activating specific genes.

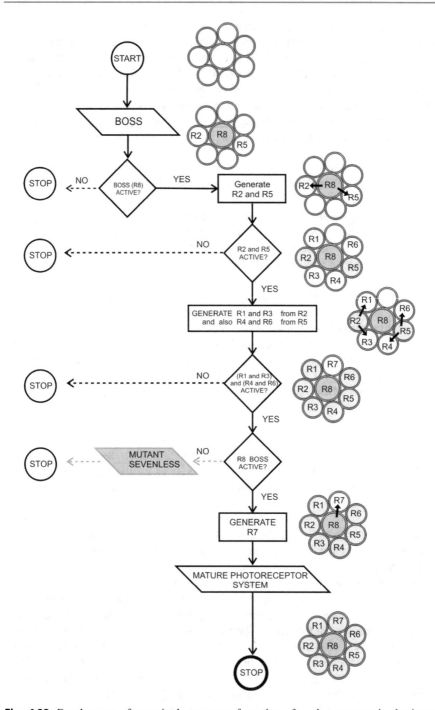

**Fig. 4.32** Development of organized structures: formation of a photoreceptor in the insect ommatidium—simplified view (active cells shown in color) and algorithmic depiction. Symbols: circle, beginning and end of the process; rectangle, command to be executed; rhombus, conditional instruction with two possible outcomes (TRUE and FALSE); parallelogram, input/output interface

## 4.7    Regulation Levels

Assuming that biological systems are, in fact, automatons and that their properties can be explained by their automatic nature, we can generalize and categorize regulatory phenomena. One example of such categorization involves treating negative feedback loops as a basic mechanism of automatic regulation and—at the same time—a basic building block of biological systems. Our ever-deepening knowledge of regulatory mechanisms seems to support this theory and provides much evidence of the universality of negative feedback loops.

The above assumptions force us to consider the function of more advanced biological mechanisms such as the nervous system, which enables fine-tuned regulation that would otherwise be impossible in a simple feedback loop. Two questions arise in this scope: are such advanced mechanisms merely modified feedback loops? How should the term "advanced regulation" be defined?

In general, an advanced regulatory system is a system where the effector has significant freedom in choosing the strategy which it will apply to a given task. In the above-described systems, effector units (particularly intracellular ones) have no such freedom—they follow a genetically determined procedure. Advanced feedback loops may, however, emerge as a result of linking many simpler systems. There are several ways in which regulatory systems may be coupled to one another. Cooperation occurs when the product of one system is used by another system. Bringing together two contradictory systems reduces the amplitude of fluctuations and promotes signal attenuation. Finally, as mostly promising for the progress, systems can be linked via coordination pathways, modifying the sensitivity of their receptors.

Sensitivity-based coupling enables one system to drive the action of another and introduces a specific hierarchy of needs, similar to the hierarchy which exists between the cell and the organism. However, coordination does not directly translate into increased effector freedom as long as at least one receptor is not subject to control. New properties may only emerge in systems which are fully coupled to each other, where neither system maintains complete dominance (see Fig. 4.33 a–c). Such conditions occur naturally in the nervous system, owing to its enormous diversity of feedback loops, all interlinked and capable of affecting one another's actions. The resulting "regulatory superloop" exhibits great variability, although it still relies on a centralized memory store. The malleability and adaptability of this complex system are far greater than that in any of its individual components. The "superloop" works by implementing a single, shared task, but in doing so, it may choose from a great number of strategies, all of which lead to the same goal.

It should be noted that such "superloops" can be further integrated with one another, providing even greater effector independence.

The presented "effector freedom" criterion enabled the French philosopher and cybernetics expert Pierre de Latil to propose the following tiered structure of regulation:

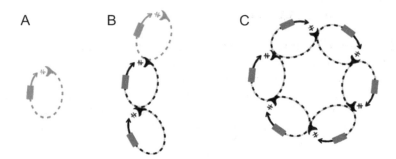

**Fig. 4.33** Coupling of regulatory loops via mutual control—formation of a "superloop"—hypothetical mechanism for the emergence of higher-level regulatory systems

1. Stabilization tier, associated with intracellular and organic regulation not mediated by the nervous system. Systems belonging to this tier work by following simple, rigid programming in order to provide answers to questions such as *how?* and *what?* For instance, if the concentration of some substance decreases, the cell must try to restore it in a predetermined way. As long as regulation is based on the action of individual circuits (regardless of their complexity), the overall system belongs to the stabilization tier.
2. Determinism tier, where systems achieve certain freedom in choosing how to implement the given tasks. This type of action is based on instincts (single answer to the *what?* question) but may involve numerous answers to the *how?* question. For instance, the need to acquire food is instinct-driven, but does not involve a single, predetermined strategy—the regulatory system exercises its freedom by choosing one of the available food acquisition strategies.
3. Goal tier, which results from development of the cerebral cortex along with massive improvements in both memory and reasoning mechanisms. Conscious, abstract thinking introduces a new quality called the goal of action, where systems may freely determine not only *how* to proceed but also *what* to do. This level of regulation is characteristic of *Homo sapiens*.

## 4.8    Hypothesis

### 4.8.1    Proteome Construction Hypothesis

Genetics and genomics (i.e., in silico genetics) have enabled scientists to study specific genes with the help of ever more accurate stochastic mechanisms. It appears that the theoretically estimated number of proteins is greater than the number of proteins actually known to biochemists. Genomics can be treated as a tool which allows us to establish a complete set of proteins, including those which have not yet been experimentally observed.

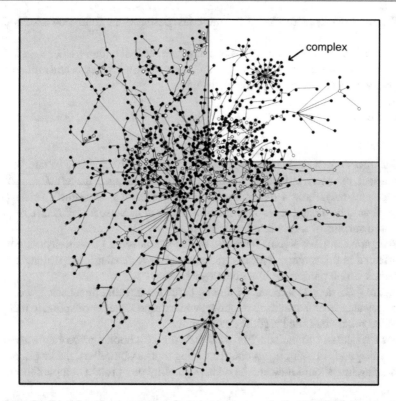

**Fig. 4.34** Traditional proteome, where each circle represents a single protein. The length of each vertex should, in theory, correspond to the duration of interaction between two proteins

Simply knowing the sequences, structures, or even functions of individual proteins does not provide sufficient insight into the biological machinery of living organisms. The complexity of individual cells and entire organisms calls for functional classification of proteins. This task can be accomplished with a *proteome*—a theoretical construct where individual elements (proteins) are grouped in a way which acknowledges their mutual interactions and interdependencies, characterizing the information pathways in a complex organism.

Most ongoing proteome construction projects focus on individual proteins as the basic building blocks (Fig. 4.34). Due to the relatively large number of proteins (between 25,000 and 40,000 in the human organism), presenting them all on a single graph with vertex lengths corresponding to the relative duration of interactions would be unfeasible. This is why proteomes are often subdivided into functional subgroups such as the metabolome (proteins involved in metabolic processes), interactome (complex-forming proteins), kinomes (proteins which belong to the kinase family), etc.

Figure 4.34 presents a sample proteome based on the assumptions and suggestions outlined in this handbook. Our model should be considered strictly hypothetical since it lacks numerical verification.

The proposed construction of the proteome bases on the following assumptions:

1. The basic unit of the proteome is one negative feedback loop (rather than a single protein) as the main regulation mechanism ensuring steady state and homeostasis.
2. Relations between units can be mediated by:
   A. Effectors
   B. Receptors


The effector relation is inherently cooperative: each effector may be targeted by a signal sent by another feedback loop. The signal changes the efficiency of the effector and may force a reaction without that effector's "consent." This type of signal should be treated as a cooperation mechanism because it is issued in response to a local surplus of a certain effector substrate.

Receptor-mediated signals force changes in the receptor's sensitivity. Any receptor affected in this manner will, in turn, adjust its own feedback loop, along with all processes which participate in a given cycle.

Figure 4.35 depicts a model proteome based on negative feedback loops. Solid lines indicate effector-mediated relations, while dashed lines correspond to receptor-mediated relations. Line length is irrelevant.

For the sake of comparison, Fig. 4.34 presents a traditional proteome, where each unit corresponds to a single protein. According to its assumptions, the length of each connecting line should indicate the relative stability of a protein complex (however,

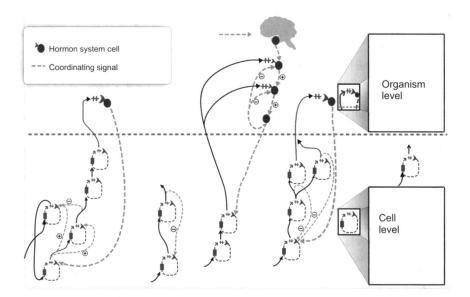

**Fig. 4.35** Model proteome conforming to the description proposed in this chapter. Negative feedback loops are represented by lines (solid lines for effector-mediated relations and dashed lines for receptor-mediated relations)

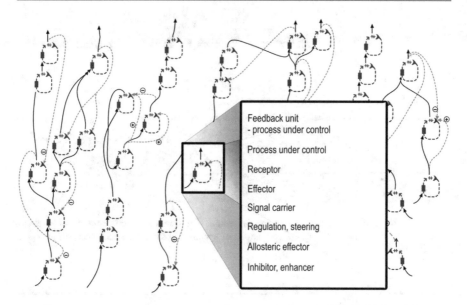

**Fig. 4.36** Proteome model (see Fig. 4.35) extended with information regarding the molecular structure of each loop

this requirement is difficult to fulfill given a large number of proteins). The model of proteome building proposed in this chapter tries to avoid taking into account individual proteins. Assuming automatic control of all biological processes and steady state as the incoming program of their action, the proteome used as a tool in studies of the cell and organism strategy could be presented as the relation of independent automatic devices rather than individual proteins. The automatic character of devices controlling biological processes allows treating them as independent units involving the final, controlled product intermediates and the whole regulatory arrangement as well as non-protein components of the process.

We believe that the construction of such proteome could reveal in future studies the still hidden strategic relations in living cells and organisms (Figs. 4.36 and 4.37).

Such a system, if properly simulated, would present a valuable tool, enabling scientists to further study the variations (both random and targeted) in biological processes. The effects of these variations could then be compared with existing medical knowledge—for example, the symptoms of known diseases. A suitable proteome model would be very helpful in ascertaining the causes and effects of complex pathological processes.

A complete, multi-tiered (including the tissue and organism tiers) in silico proteome might even be called an "*in silico* organism"—it would facilitate virtual experiments on humans which cannot be carried out in the real world for technical or ethical reasons. It would also provide valuable insight into systemic pathologies which are difficult to study when focusing on small-scale subunits of living organisms, such as individual cells.

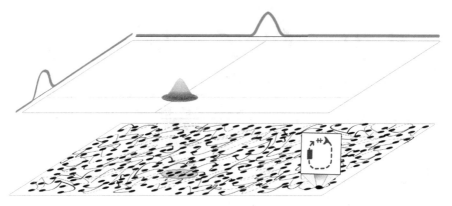

**Fig. 4.37** The multi-tier properties of the proteome. The bottom tier corresponds to feedback loops, while the top tier involves concentrations of specific substances in each compartment of a fully differentiated cell

## Suggested Reading

Austad SN, Finch CE (2022) How ubiquitous is aging in vertebrates? Science 376(6600): 1384–1385. https://doi.org/10.1126/science.adc9442

Bays JL, Campbell HK, Heidema C, Sebbagh M, DeMali KA (2017) Linking E-cadherin mechanotransduction to cell metabolism through force-mediated activation of AMPK. Nat Cell Biol 19(6):724–731. https://doi.org/10.1038/ncb3537

Beatty J, Legewie H (1976) Biofeedback of behavior NATO Conference Series Serie III – human factors. Plenum Press, New York

Boucard AA (2022) Self-activated adhesion receptor proteins visualized. Nature 604:628–630. https://doi.org/10.1038/d41586-022-00972-0

Bowie JU (2005) Solving the membrane protein folding problem. Nature 438(7068):581–589. https://doi.org/10.1038/nature04395

Brodsky IE (2020) Caspase-8 protein cuts a brake on immune defences. Nature 587(7833): 201–203. https://doi.org/10.1038/d41586-020-02994-y

Buchman TG, Ferrel JE, Li R, Meyer T (2005) Interlinked fast and slow positive feedback loops drive reliable cell decisions. Science 310:496–498

Changeux J-P, Edelstein SJ (2005) Allosteric mechanisms of signal transduction. Science 308: 1424–1428

Cheng Y (2018) Single-particle cryo-EM-How did it get here and where will it go. Science 361(6405):876–880. https://doi.org/10.1126/science.aat4346

Collier A, Liu A, Torkelson J, Pattison J, Gaddam S, Zhen H, Patel T, McCarthy K, Ghanim H, Oro AE (2022) Gibbin mesodermal regulation patterns epithelial development. Nature 606(7912): 188–196. https://doi.org/10.1038/s41586-022-04727-9

Corbett EF, Michalak M (2000) Calcium, a signalling molecule in the endoplasmic reticulum? TIBS 25:307–310

Dequeker BJH, Scherr MJ, Brandão HB, Gassler J, Powell S, Gaspar I, Flyamer IM, Lalic A, Tang W, Stocsits R, Davidson IF, Peters J-M, Duderstadt KE, Mirny LA, Tachibana K (2022) MCM complexes are barriers that restrict cohesin-mediated loop extrusion. Nature 606(7912): 197–203. https://doi.org/10.1038/s41586-022-04730-0

Ding Y, Tang Y, Kwok CK, Zhang Y, Bevilacqua PC, Assmann SM (2014) In vivo genome-wide profiling of RNA secondary structure reveals novel regulatory features. Nature 505(7485): 696–700. https://doi.org/10.1038/nature12756

Essuman K, Milbrandt J, Dangl JL, Nishimura MT (2022) Shared TIR enzymatic functions regulate cell death and immunity across the tree of life. Science 377(6605):eabo0001. https://doi.org/10.1126/science.abo0001

Ferrell JE (2002) Self-perpetuating states in signal transduction: positive feedback, double-negative feedback and bistability. Curr Opin Chem Biol 6:140–148

Floyd BJ, Wilkerson EM, Veling MT, Minogue CE, Xia C, Beebe ET, Wrobel RL, Cho H, Kremer LS, Alston CL, Gromek KA, Dolan BK, Ulbrich A, Stefely JA, Bohl SL, Werner KM, Jochem A, Westphall MS, Rensvold JW, Taylor RW, Prokisch H, Kim JP, Coon JJ, Pagliarini DJ (2016) Mitochondrial protein interaction mapping identifies regulators of respiratory chain function. Mol Cell 63(4):621–632. https://doi.org/10.1016/j.molcel.2016.06.033

Gao XJ, Chong LS, Kim MS, Elowitz MB (2018) Programmable protein circuits in living cells. Science 361(6408):1252–1258. https://doi.org/10.1126/science.aat5062

Goutam K, Ielasi FS, Pardon E, Steyaert J, Reyes N (2022) Structural basis of sodium-dependent bile salt uptake into the liver. Nature 606(7916):1015–1020. https://doi.org/10.1038/s41586-022-04723-z

Guo Z, Ohlstein B (2015) Stem cell regulation. Bidirectional Notch signaling regulates Drosophila intestinal stem cell multipotency. Science 350(6263):aab0988. https://doi.org/10.1126/science.aab0988

Hernández AR, Klein AM, Kirschner MW (2012) Kinetic responses of β-catenin specify the sites of Wnt control. Science 338:1337–1340

Hodges A (2012) Beyond Turing's machines. Science 336:163–164

Holmström KM, Finkel T (2014) Cellular mechanisms and physiological consequences of redox-dependent signalling. Nat Rev Mol Cell Biol 15(6):411–421. https://doi.org/10.1038/nrm3801

Hu H, Juvekar A, Lyssiotis CA, Lien EC, Albeck JG, Oh D, Varma G, Hung YP, Ullas S, Lauring J, Seth P, Lundquist MR, Tolan DR, Grant AK, Needleman DJ, Asara JM, Cantley LC, Wulf GM (2016) Phosphoinositide 3-kinase regulates glycolysis through mobilization of aldolase from the actin cytoskeleton. Cell 164(3):433–446. https://doi.org/10.1016/j.cell.2015.12.042

Itoh Y, Khawaja A, Laptev I, Cipullo M, Atanassov I, Sergiev P, Rorbach J, Amunts A (2022) Mechanism of mitoribosomal small subunit biogenesis and preinitiation. Nature 606(7914): 603–608. https://doi.org/10.1038/s41586-022-04795-x

Kaplan T, Friedman N (2012) Running to stand still. Nature 484:171–172

Kornberg RD (1996) RNA polymerase II transcription control. TIBS 21:325–326

de Latil P (1953) La pensée artificielle. Ed. Gallimard, Paris

Li F, Eriksen J, Finer-Moore J, Chang R, Nguyen P, Bowen A, Myasnikov A, Yu Z, Bulkley D, Cheng Y, Edwards RH, Stroud RM (2020) Ion transport and regulation in a synaptic vesicle glutamate transporter. Science 368(6493):893–897. https://doi.org/10.1126/science.aba9202

Lisman JE (1985) A mechanism for memory storage insensitive to molecular turnover: a bistable autophosphorylating kinase. Proc Natl Acad Sci USA 82:3055–3057

Mamou G, Corona F, Cohen-Khait R, Housden NG, Yeung V, Sun D, Sridhar P, Pazos M, Knowles TJ, Kleanthous C, Vollmer W (2022) Peptidoglycan maturation controls outer membrane protein assembly. Nature 606(7916):953–959. https://doi.org/10.1038/s41586-022-04834-7

Mayer B, Hemmens B (1997) Biosynthesis and action of nitric oxide in mammalian cells. TIBS 22: 477–481

Meldolesi J, Pozzan T (1998) The endoplasmic reticulum $Ca^+$ store: a view from the lumen. TIBS 23:10–14

Mitchell L, Chang G, Horton NC, Kercher MA, Pace HC, Schumacher MA, Brennan R, Lu GP (1996) Crystal structure of the lactose operon repressor and its complexes with DNA and inducer. Science 271:1247–1254

Mottis A, Herzig S, Auwerx J (2019) Mitocellular communication: shaping health and disease. Science 366(6467):827–832. https://doi.org/10.1126/science.aax3768

Naik S, Fuchs E (2022) Inflammatory memory and tissue adaptation in sickness and in health. Nature 607(7918):249–255. https://doi.org/10.1038/s41586-022-04919-3

Neumayr C, Haberle V, Serebreni L, Karner K, Hendy O, Boija A, Henninger JE, Li CH, Stejskal K, Lin G, Bergauer K, Pagani M, Rath M, Mechtler K, Arnold CD, Stark A (2022) Differential cofactor dependencies define distinct types of human enhancers. Nature 606(7913): 406–413. https://doi.org/10.1038/s41586-022-04779-x

Nicholson WD, Thornberry NA (1997) Caspases: killer proteases. TIBS 22:299–306

Nozaki K, Maltez VI, Rayamajhi M, Tubbs AL, Mitchell JE, Lacey CA, Harvest CK, Li L, Nash WT, Larson HN, McGlaughon BD, Moorman NJ, Brown MG, Whitmire JK, Miao EA (2022) Caspase-7 activates ASM to repair gasdermin and perforin pores. Nature 606(7916):960–967. https://doi.org/10.1038/s41586-022-04825-8

Park JS, Burckhardt CJ, Lazcano R, Solis LM, Isogai T, Li L, Chen CS, Gao B, Minna JD, Bachoo R, DeBerardinis RJ, Danuser G (2020) Mechanical regulation of glycolysis via cytoskeleton architecture. Nature 578(7796):621–626. https://doi.org/10.1038/s41586-020-1998-1

Pines J (1999) Four-dimensional control of the cell cycle. Nat Cell Biol 1:E73–E79

Qing Y, Ionescu SA, Pulcu GS, Bayley H (2018) Directional control of a processive molecular hopper. Science 361(6405):908–912. https://doi.org/10.1126/science.aat3872

Rensvold JW, Shishkova E, Sverchkov Y, Miller IJ, Cetinkaya A, Pyle A, Manicki M, Brademan DR, Alanay Y, Raiman J, Jochem A, Hutchins PD, Peters SR, Linke V, Overmyer KA, Salome AZ, Hebert AS, Vincent CE, Kwiecien NW, Rush MJP, Westphall MS, Craven M, Akarsu NA, Taylor RW, Coon JJ, Pagliarini DJ (2022) Defining mitochondrial protein functions through deep multiomic profiling. Nature 606(7913):382–388. https://doi.org/10.1038/s41586-022-04765-3

Rosenfeld N, Young JW, Alon U, Swain PS, Elowitz MB (2005) Gene regulation at the single-cell level. Science 307:1962–1969

Scheffzek K, Ahmadian MR, Wittinghofer A (1998) GTPase-activating proteins: helping hands to complement an active site. TIBS 23:257–262

Takano T, Wallace JT, Baldwin KT, Purkey AM, Uezu A, Courtland JL, Soderblom EJ, Shimogori T, Maness PF, Eroglu C, Soderling SH (2020) Chemico-genetic discovery of astrocytic control of inhibition in vivo. Nature 588:297–302. https://doi.org/10.1038/s41586-020-2926-0

Tanner LB, Goglia AG, Wei MH, Sehgal T, Parsons LR, Park JO, White E, Toettcher JE, Rabinowitz JD (2018) Four key steps control glycolytic flux in mammalian cells. Cell Syst 7(1):49–62.e8. https://doi.org/10.1016/j.cels.2018.06.003

Thomas R, Kauffman M (2001) Multistationarity, the basis of cell differentiation and memory. I. Structural conditions of multistationarity and other nontrivial behaviour. Chaos 11:170–178

Vanhaesebroeck B, Leevers SJ, Panayotou G, Waterfield MD (1997) Phosphoinositide-3 kinases: a conserved family of signal transducers. TIBS 22:267–272

Veerakumar A, Yung AR, Liu Y, Krasnow MA (2022) Molecularly defined circuits for cardiovascular and cardiopulmonary control. Nature 606(7915):739–746. https://doi.org/10.1038/s41586-022-04760-8

Verdin E (2015) NAD$^+$ in aging, metabolism, and neurodegeneration. Science 350(6265): 1208–1213. https://doi.org/10.1126/science.aac4854

Voet D, Voet JG, Pratt CW (1999) Fundamentals of biochemistry. Wiley, New York

Wall ME, Dunlop MJ, Hlavacek WS (2005) Multiple functions of a feed-forward-loop gene circuit. J Mol Biol 349:501–514

Wallach D (1997) Cell death induction by TNF: a matter of self-control. TIBS 22:107–109

Wang QC, Zheng Q, Tan H, Zhang B, Li X, Yang Y, Yu J, Liu Y, Chai H, Wang X, Sun Z, Wang JQ, Zhu S, Wang F, Yang M, Guo C, Wang H, Zheng Q, Li Y, Chen Q, Zhou A, Tang TS (2016) TMCO1 is an ER Ca(2+) load-activated Ca(2+) channel. Cell 165(6):1454–1466. https://doi.org/10.1016/j.cell.2016.04.051

De Witt MA, Chang AY, Combs PA, Yildiz A (2012) Cytoplasmic dynein moves through uncoordinated stepping of the AAA+ ring domains. Science 335:221–225

Wittinghofer A, Nassar N (1996) How RAS-related proteins talk to their effectors. TIBS 21:488–491

Xiao B, Sanders MJ, Underwood E, Heath R, Mayer FV, Carmena D, Jing C, Walker PA, Eccleston JF, Haire LF, Saiu P, Howell SA, Aasland R, Martin SR, Carling D, Gambin SJ (2011) Structure of mammalian AMPK and its regulation by ADP. Nature 472:230–233

Xu D, Zhao H, Jin M, Zhu H, Shan B, Geng J, Dziedzic SA, Amin P, Mifflin L, Naito MG, Najafov A, Xing J, Yan L, Liu J, Qin Y, Hu X, Wang H, Zhang M, Manuel VJ, Tan L, He Z, Sun ZJ, Lee VMY, Wagner G, Yuan J (2020) Modulating TRADD to restore cellular homeostasis and inhibit apoptosis. Nature 587(7832):133–138. https://doi.org/10.1038/s41586-020-2757-z

Yalcin A, Telang S, Clem B, Chesney J (2009) Regulation of glucose metabolism by 6-phosphofructo-2-kinase/fructose-2,6-bisphosphatases in cancer. Exp Mol Pathol 86(3):174–179. https://doi.org/10.1016/j.yexmp.2009.01.003

Yang W, Hekimi S (2010) A mitochondrial superoxide signal triggers increased longevity in Caenorhabditis elegans. PLoS Biol 8(12):e1000556. https://doi.org/10.1371/journal.pbio.1000556

Ye Y, Blaser G, Horrocks MH, Ruedas-Rama MJ, Ibrahim S, Zhukov AA, Orte A, Klenerman D, Jackson SE, Komander D (2012) Ubiquitin chain conformation regulates recognition and activity of interacting proteins. Nature 492:266–270

Zoladz JA, Duda K, Majerczak J (1998) Gas exchange, blond acid-base balance and mechanical muscle efficiency during incremental levels of exertion In young healthy individuals. Pneumonol Alergol Pol 66(3–4):163–172

# Interrelationship in Organized Biological Systems

**5**

*Interrelationship is a prerequisite of function in organized structures:*

## Abstract

Ensuring synchronization between processes which make up a complex system requires functional interdependence, with multiple mechanisms cooperating to achieve a predetermined final state. **Interrelationship involves multiple systems, each equipped with its own regulatory mechanism and hence operating as a distinct entity**. Such structural and functional isolation is evident in many complex systems. Mutual dependencies may take on two forms: **cooperation and coordination**. If the product of one process acts as a substrate for another, we are dealing with cooperation. In such cases both processes are subject to their own separate control mechanisms. Coordination occurs when one process acts directly upon the control mechanisms of other processes, thereby establishing hierarchical dependence. Biological hierarchies arise due to allosteric properties of receptor units which can register signals issued by other autonomous processes. Compartmentalization of mutually dependent processes optimizes the functioning of cellular machinery.

## Keywords

Cooperation and coordination · Organization of interrelationship · Coordination strategies · Signaling · Role of hormonal signals · Intracellular signaling

1. The diversity of cooperation mechanisms in cellular and organic metabolism.
2. The diversity of coordination mechanisms in cellular and organic metabolism.
3. Hepatic cell activity during periods of energy abundance.
4. Hepatic cell activity during periods of energy deficiency.
5. Co-action in distress—acute shortage of oxygen or glucose.
6. Why do cells contain organelles?
7. The properties of regulatory hierarchy (organism vs. cell).
8. Coordination of metabolism and the role of biological clocks.
9. Pathologies resulting from disruption of co-action (e.g., tumors).
10. Glucose levels as the coordinating factor in energy management—properties.
11. Why can an anthill be referred to as an organism?
12. What are the consequences of glycolysis—a key aspect of energy management—occurring outside of the mitochondrion?
13. How is co-action between hepatic and muscle tissues realized?
14. How are programs of action encoded in biology?

## 5.1    The Need of Mutual Relations in Biological Systems

Assuming that all structures involved in biological functions belong to regulatory loops allows us to determine—even considering the limitations of our scientific knowledge—the principles of regulation and mutual relations in such automatic systems (as opposed to non-automatic systems, subjected to external decision-making mechanisms and therefore unpredictable when acting on their own). Automaticity is tied to independence, which can only be achieved in the presence of a negative feedback loop. Individual components of an automatic system cannot be considered independent.

As processes owe their independence to automatic control mechanisms, they need to cooperate with one another to fulfill specific biological goals.

Stabilization of substance concentrations and/or biological activity is a result of genetic programming implemented by the cell. In an ideal organism or cell existing in an unchanging environment, all forms of cooperation other than the exchange of reaction substrates could be neglected (in particular, there would be no need for activation or inhibition mechanisms). In reality, however, interaction with the external environment introduces random stimuli which translate into unpredictable fluctuations in the intensity of biological processes and the availability of key substances. Such fluctuations can be caused by inflow and outflow of substrates and reaction products as well as by forced activation of cells. The problem is compounded by the existence of a barrier (membrane) which hampers transmission of substances and signals between the cell and the interstitial space. Unpredictability becomes particularly troublesome in processes whose individual stages occur in different areas of the cell.

Significant deviations from the norm may arise as a result of variable intensity of biological processes. While oscillations are an inherent feature of systems subjected to automatic regulation with negative feedback loops, they may become a problem if the resulting changes are too severe.

## 5.2 Cooperation and Coordination

Regulatory systems may assist one another either by cooperation or by coordination.

Cooperation is understood as correlation of independent processes based on a shared pool of products or substrates, where neither process is subordinated to the other. Coordination is also a manifestation of correlated activities, but in this case one process is in direct control of the other, giving rise to a hierarchical structure. Biological cooperation may be compared to a contract between two factories, one of which makes ball bearings, while the other uses them in the machines it assembles. In a cooperative system, both factories retain their independence. If, however, the ball bearing manufacturer were to become a branch of the machine factory or if its production rate was fully dependent on the machine assembly rate, we would be dealing with more than a simple sale/purchase contract. In such circumstances, control over the entire manufacturing process would rest with the machine factory. Clearly, such centralized control introduces a hierarchical structure and should therefore be treated as a form of coordination.

Similar properties can be observed in biological systems. While the "technicalities" of coordination and cooperation mechanisms differ from process to process, their general principles remain the same. Understanding them provides insight into the structure and function of biological systems.

The nature of a negative feedback loop, which underpins biological regulatory systems, is to restore stability. Maintaining a steady state is therefore—out of necessity—the core principle of biological programming, both in the cell and in the organism as a whole.

Major deviations from the evolutionarily conditioned substance or activity levels are caused by uncontrollable, random events originating in the external environment. This applies both to cells and to organisms. Individual cells operate in a state of relative homeostasis, but they also participate in hierarchical systems, and their function is subordinated to the requirements of their host organism. The organism can force activation or deactivation of certain cells in order to stabilize its own vital parameters, such as blood glucose levels. The signals issued by the organism (hormones) are treated as commands and may activate various mechanisms in each cell for the benefit of a distributed process, producing results which the cell would not otherwise generate on its own.

The organism comes into frequent contact with the outside world; for instance, through variations in food intake, however, such interaction presents dangers for the state of homeostasis.

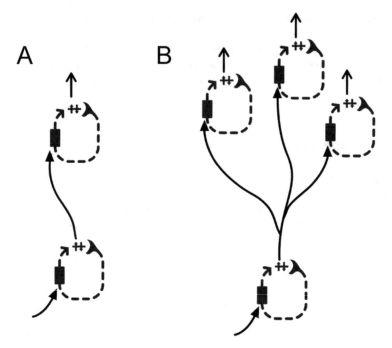

**Fig. 5.1** The principle of cooperation between processes subjected to their own negative feedback loops. (**a**) The product of one process used as a substrate in the effector unit of another process, in accordance with a negative feedback loop. (**b**) The product of one process used as a substrate for many other processes (e.g., glucose-6-phosphate synthesis providing substrates for the pentose cycle, glycolysis, glycogen synthesis, etc.)

Due to the expected randomness and fluctuations, biological systems cannot rely solely on cooperation. The organism responds to these challenges by coordinating its internal processes via hormonal signals.

In order to perform the action requested by the organism, the cell must first ensure its own stability by means of internal coordination. It does so by generating further signals which guide its internal processes. This property allows the cell to counteract major deviations which emerge as a result of inconsistent activity of internal regulatory mechanisms.

In individual cells, both cooperation and coordination are enhanced by the limited space in which the linked processes need to take place.

Substance exchange is the most basic form of cooperation. As such, it is frequently observed in living cells. Examples include the coupling between DNA synthesis and the pentose cycle (which provides pentoses), between glycogen synthesis and glucose-6-phosphate synthesis, between globin synthesis and heme synthesis, etc. Such cooperation occurs when the product of one reaction is a substrate for another reaction, resulting in a "supply chain" condition. Increased demand for the final product causes a decrease in its concentration. This, in turn, triggers control mechanisms which act to increase supply. Figures 5.1, 5.2, and 5.3

**Fig. 5.2** Cooperation as
applied to cellular metabolism

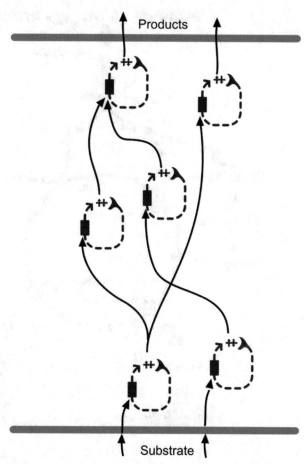

present the cooperation of automatic processes where the product of one process is used as a substrate in the effector unit of another process.

If, however, the cooperation of two processes is subject to significant disruptions (for instance, if both processes occur in separate compartments of the cell or if the rapid reaction rate makes storage of intermediate products unfeasible), coordination becomes a necessity. Coordinating signals act as "administrators," increasing demand for overproduced substances or inhibiting the processes which produce them. Coordination is usually implemented by exploiting the product of one reaction as an allosteric effector (rather than a substrate) in another process. The controlling process modifies the sensitivity of the receptor loop instead of directly altering the rate at which a given product is consumed. Figure 5.4 presents a model view of coordination.

In systems which rely on negative feedback loops, the stabilization curve typically assumes the form of a sinusoid. The effects of cooperation and coordination can be described as changes in its shape or oscillation level.

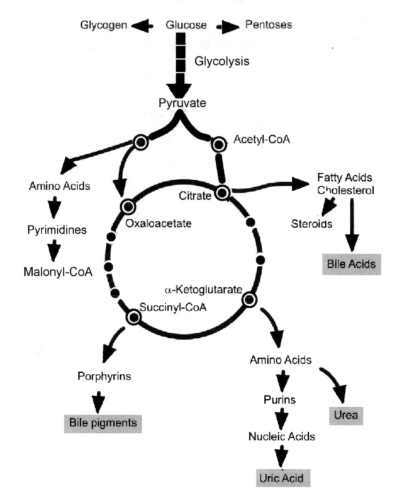

**Fig. 5.3** Cooperative alignment of the Krebs cycle with other biological processes

The action of a cooperating system which consumes a given substance may result in changed frequency and (optionally) amplitude of oscillations of its stabilization curve. Increased demand for the product precipitates a decline in its concentration and triggers an increase in its production rate. However, as the receptors of cooperating systems do not change their sensitivity, the target product concentration must remain the same. Thus, any changes are effectively limited to the shape of the stabilization curve (see Fig. 5.5). On the other hand, coordinating signals (action of allosteric effectors) alter the sensitivity of receptors, thereby modifying the initial biological program and forcing a change in the target product concentration or activity level (Fig. 5.6). This phenomenon manifests itself as a change in the slope gradient on the attached allosteric receptor transformation diagram (Fig. 5.7),

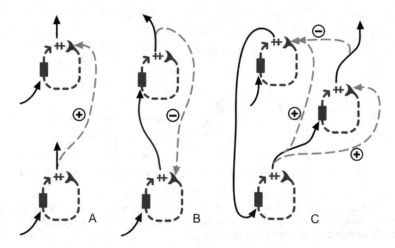

**Fig. 5.4** Model view of coordination. The product of one process affects other processes which are also subject to their own, independent negative feedback loops (the coordinating product downregulates the receptor by reducing its affinity to its own product). (**a**) Coordination of indirectly coupled processes (without direct cooperation) and (**b** and **c**) coordination of directly coupled processes (with direct cooperation). Solid lines indicate cooperative links. Dashed (colored) lines indicate coordinating signals

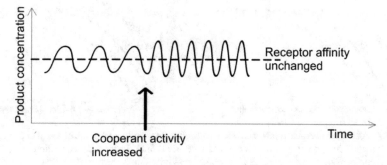

**Fig. 5.5** Change in the frequency of the product stabilization curve in a self-controlled process, resulting from increased activity of a cooperating system (increased demand for the product)—an example of product concentration variability

similarly to the effect of allosteric effectors on hemoglobin. It can be observed both in the covalent modification of enzymes (i.e., the action of hormones) and as a result of noncovalent interaction initiated by an allosteric effector (e.g., an intracellular coordinating signal).

In general, coordination can be defined as any action which affects the sensitivity of receptors other than one's own, facilitating interrelation of separate systems in the pursuit of a common goal.

As both regulatory and coordinating mechanisms exploit the allosteric properties of signal-receptor interactions, distinguishing them may appear difficult. We can,

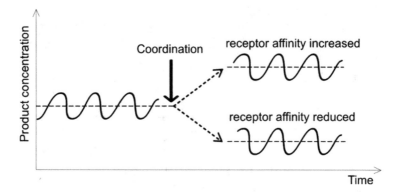

**Fig. 5.6** Change in the target concentration as an effect of coordinating processes (decreased receptor sensitivity results in increased concentration of the product, while an increase in sensitivity causes the product concentration to decrease)

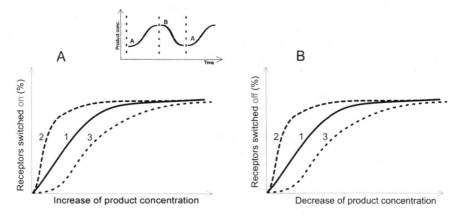

**Fig. 5.7** Allosteric receptor transformation curve: changes resulting from the effect of coordinating signals (dashed line) compared with the initial state (solid line). Product concentration increases in order to compensate for the decrease in receptor sensitivity (and vice versa)

however, assume that any loop in which the detector subunit exhibits affinity to the product released by its corresponding effector is a regulatory mechanism. If the detector also registers signals from other systems, the changes triggered by such signals may be counted among coordinating effects (Figs. 5.8 and 5.9). For instance, if a blood glucose concentration receptor triggers a process which results in modification of blood glucose levels, the action of the system can be best described as self-regulatory. If, however, the same detector also activates other processes, resulting, e.g., in modifying the concentration of fatty acids in blood, then we are dealing with coordination.

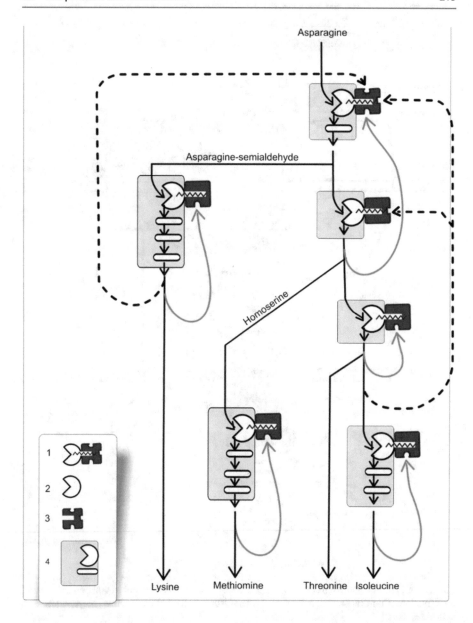

**Fig. 5.8** Control (regulation, colored line) of coupled processes via cooperation (solid line) and coordination (dashed line), as applied to amino acid synthesis. Each fork in the synthesis pathway involves separate coordinating mechanisms. Symbols are explained in the inset. Symbolic expressions: 1, regulatory enzyme; 2, catalytic subunit; 3, regulatory (receptor) subunit; 4, effector and its enzymes

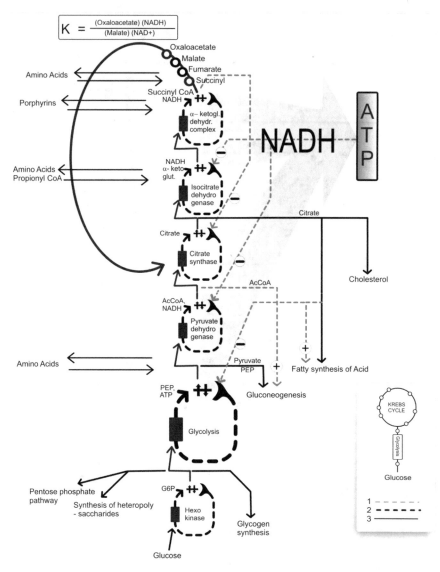

**Fig. 5.9** Glycolysis and Krebs cycle divided into stages, with focus on their regulation and mutual coordination. Coordinating signals, line 1; regulatory signals, line 2; flow of products and substrates, line 3

## 5.3    The Characteristics of Process Coordination in Individual Cells and Organisms

Intracellular coordinating signals may be treated as counterparts of hormonal signals: they both share a similar mechanism of action, acting on receptors and altering their most fundamental property, i.e., sensitivity. However, the specificity of intracellular coordination pathways is somewhat different than in the case of hormones: such pathways rely on noncovalent interaction between signals and receptors (contrary to hormones, which usually trigger phosphorylation or other covalent modifications). It can therefore be said that intracellular coordinating signals exert *influence* on independent processes, but do not directly *command* them.

Another difference between both types of signals is the principle of unamplified action. Inside a cell, the distance traveled by a coordinating signal is usually short (negligible dilution of signal), negating the need for amplification. This issue is also related to the lack of a need for encoding. Encoding systems become necessary when the coordinating signal must undergo significant amplification—a process which also happens to aggravate any errors or distortions. The cell signals which effect intracellular coordination processes (products of the regulatory loops) are commonly called allosteric effectors.

Although the presented principles seem very generic, biological systems sometimes depart from them. Such departures involve both intracellular and hormonal signals (i.e., signals which the organism uses to control the function of individual cells). They are usually observed in circumstances where following established principles would result in unacceptable risk to proper transduction of the signal or the action of any component of the feedback loop (effector or receptor).

As a general rule, hormonal signals override intracellular signals. Hormones act by forming permanent, covalent bonds with proteins mediating signal and should therefore be treated as *commands*. However, in some cases obeying such commands involves the risk that the cell will become exhausted and unable to maintain its own vital parameters at an acceptable level. This issue is particularly important for enzymes which interfere with key cellular machinery. One example is the action of glucagon and insulin which control a critically important process (glycolysis) and can therefore easily drive the cell into metabolic distress. Both hormones strongly affect glycolysis by way of phosphofructokinase phosphorylation; however instead of directly interacting with phosphofructokinase-1, they act upon phosphofructokinase-2—an enzyme whose product forms a noncovalent bond with phosphofructokinase-1. Owing to this mechanism, a critically stressed hepatocyte may refuse to obey a command which threatens its own survival. This phenomenon is, however, something of an exception: similar safeguards do not exist in many other less important for cell survival processes, for instance, glycogen degradation, where the enzyme acting directly on glycogen (glycogen phosphorylase) is subject to covalent modification by hormones.

Atypical properties can also be observed in certain intracellular pathways such as the urea cycle, which involves signal coding. As noted above, intracellular signals are not usually encoded because they do not require amplification and are therefore

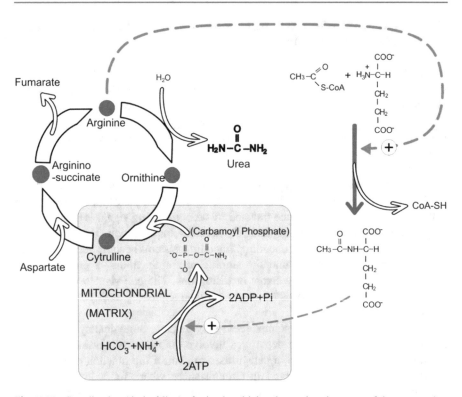

**Fig. 5.10** Coordination (dashed line) of mitochondrial and cytoplasmic stages of the urea cycle

largely protected against errors. Encoding may, however, become necessary whenever there is a significant risk of signal misinterpretation. Such a situation is observed in the urea synthesis process where an intense influx of nitric substrates into the hepatic cell may be uncorrelated with mitochondrial synthesis of carbamoyl phosphate.

As amino acid components of the urea cycle may be found in the mitochondrion regardless of this metabolic pathway, there is a risk that the coordinating signal could be misinterpreted if the messenger molecule were a simple amino acid. Hence, the urea cycle coordinator must be based on a substance which does not normally occur in mitochondria. This requirement is fulfilled by N-acetylglutamate which indicates the concentration of arginine in the cytoplasm. Synthesis of N-acetylglutamate is triggered by arginine, whose high concentration means that ornithine (a competitive arginase inhibitor) is not being transformed into citrulline at a rapid enough rate and that the process needs to be upregulated (Fig. 5.10).

We can therefore conclude that while standard signal transduction principles apply in most circumstances, special cases may require deviations from generic solutions.

## 5.4   Mutual Relation Between Cells and the Organism: Activation and Inhibition of Enzymes (Rapid Effects)

Interrelationship enables biological systems to perform advanced tasks; however it requires efficient cooperation and coordination.

The blueprints for mutual support determine strategies applied by individual cells and the whole organism in pursuit of their shared goal, i.e., homeostasis.

Cells express their specialized functions in response to commands issued by the organism. They are, however, autonomous, and if commands do not arrive, they may become independent, which is evidenced by changes in their behavioral strategies.

The mechanisms of cellular cooperation and coordination, along with their impact on energy transfer and storage processes, have been extensively studied. They can be easily traced, e.g., by observing the hormonal regulation of hepatic functions following consumption of food (fed state), during fasting and in intermediate phases. Given an abundance of nutrients (i.e., immediately following a meal), the organism releases a hormone (insulin) which forces hepatocytes to intensify their involvement in nutrient sequestration. On the other hand, if the organism is starving, liver cells are mobilized to inject additional nutrients into the bloodstream by consuming energy stores.

Regulation and coordination mechanisms involved in day-to-day functioning of the organism must be able to react to stimuli in a timely fashion. Thus, energy transfer and storage should be managed by activation and inhibition of existing enzymes rather than by changing their concentrations via transcription and translation.

Figure 5.11 presents the effect of insulin on liver cells—promoting sequestration of nutrients following a meal. In this case, the hepatocyte synthesizes fatty acids by activating carboxylase, acetyl-CoA, and ATP citrate lyases. Sequestration of triglycerides in the form of adipose tissue is enabled by insulin-mediated absorption of glucose into adipocytes. The combined effect of high blood glucose levels and the topping-off of glycogen stores effectively prevent the organism from further synthesis of sugars (gluconeogenesis).

Large-scale synthesis of fatty acids in liver cells requires significant quantities of ATP and involves oxidation processes where some of the nutrients are burned in order to power ongoing reactions. Thus, fatty acid synthesis is coupled to synthesis of ATP and NADPH. In fact, according to regulation principles, it is the overabundance of ATP and NADH which inhibits citrate synthase, isocitrate dehydrogenase, and α-ketoglutarate dehydrogenase. The role of this mechanism is to supply sufficient energy for sequestration processes. Most of the available glucose is converted into fatty acids, while the rest is consumed in the pentose shunt associated with production of NADPH.

During periods of starvation (Fig. 5.12) the role of glucagon increases (i.e., the insulin/glucagon ratio drops). The increased release of glucagon is a result of decreased blood glucose levels (note that maintaining steady concentration of glucose in blood is of critical importance to the organism). Glucagon works by reorienting the organism's energy transfer and storage processes. Fatty acid

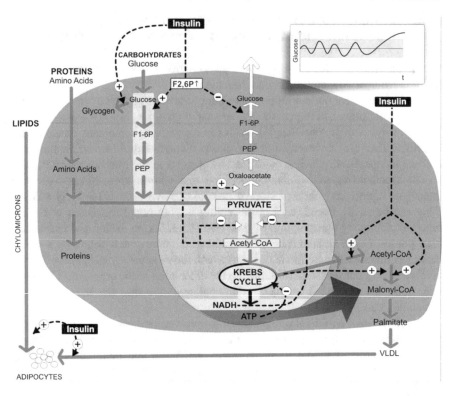

**Fig. 5.11**  Control of sequestration processes in a liver cell. Dashed lines indicate regulatory and coordinating signals. The top right-hand inset depicts fluctuations in blood glucose levels as a result of the presented activity. Upregulated metabolic pathways are shown in color

synthesis is halted via inhibition of acetyl-CoA carboxylase and pyruvate kinase as well as activation of fructose-1,6-bisphosphatase. Blood glucose deficiency is compensated for by stimulation of gluconeogenesis, i.e., synthesis of glucose from amino acids, lactate, and glycerol derived from lipolysis. When oxaloacetate becomes scarce, the liver cell may draw additional energy from β-oxidation, ejecting ketone bodies (a byproduct of acetyl-CoA) into the bloodstream.

During intermediate stages, i.e., when the blood glucose concentration is close to its biological optimum (approximately 5 mmol/l) and when insulin and glucagon are in relative equilibrium, the cell becomes independent. Its own regulatory mechanisms (based on noncovalent bonds) assume control, while the energy conversion processes occurring inside the cell are subject to allosteric regulation, primarily by ATP and NADH. The cell now focuses on maintaining optimal levels of ATP and NADH in its own cytoplasm and does not actively participate in ensuring homeostasis of the organism as a whole (Fig. 5.13).

An independent hepatocyte may synthesize fatty acids and glucose even when no hormonal signals are present; however in such cases, sequestration of nutrients is not the primary goal of the cell. Rather, the hepatocyte concerns itself with maintaining a

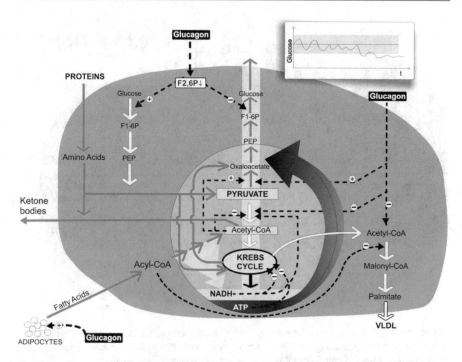

**Fig. 5.12**  Changes in hepatic metabolism resulting from reduction in blood glucose levels. The top right-hand inset depicts the fluctuating blood glucose levels. Dashed lines indicate intracellular regulatory and coordinating signals. Upregulated metabolic pathways are shown in color

steady concentration of energy carriers, treating synthesis as a convenient way to purge excess carriers. This change in strategy becomes evident in the readjustment of regulatory mechanisms involved, e.g., in glycolysis. Given an abundance of nutrients, glycolysis is activated by an external mediator (insulin) as a prerequisite of intensified sequestration. In an independent cell, the same process is controlled by citrate, which promotes fatty acid synthesis but—unlike insulin—also inhibits glycolysis. This means that a cell which is already rich in ATP may actively protect itself from further absorption of nutrients. Consequently, both ATP degradation pathways (fatty acid synthesis activated by citrate acid as well as glucose synthesis mediated by ATP and acetyl-CoA) remain open. The relatively low demand for glucose (compared with post-meal conditions) is satisfied by absorbing glucose directly from the bloodstream or by extracting it from glycogen with the use of glycogen phosphorylase, whose activity remains low but sufficient to cover the energy requirements of the liver cell.

Full independence of the hepatocyte is a short-lived condition and does not directly correspond to periods between meals (which can indeed be quite long). Cells can remain independent only when the contradictory activity of insulin and glucagon remains in equilibrium; however both hormones are active even when no

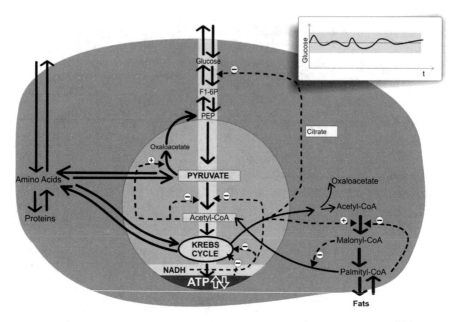

**Fig. 5.13** Regulation and coordination of energy conversion processes in a hepatic cell given a lack of coercive hormonal signals. The top right-hand inset depicts the fluctuating blood glucose levels. Dashed lines indicate intracellular regulatory and coordinating signals

food is being consumed—for instance, when strenuous physical exertion causes a significant drop in blood glucose levels.

Figures 5.14, 5.15 and 5.16 present organ collaboration in the scope of energy management, as reflected by the relation between glucagon and insulin levels. Hormone concentrations range from 0.5 $\mu U$ $pg^{-1}$ (U, unit) following a meal to 0.05 $\mu U$ $pg^{-1}$ during periods of starvation. During intermediate stages, when hepatocytes act independently, these values are close to 0.15 $\mu U$ $pg^{-1}$.

Tight collaboration between hepatocytes and the organism, regardless of the availability of nutrients (i.e., when nutrients are abundant, during periods of starvation and in intermediate stages), suggests that—according to the principles outlined above—cells and organism share similar goals and collaborate even though each of these units is subject to its own regulatory mechanisms. In both cases regulation occurs automatically, according to biological programming. Both types of regulatory loops act to stabilize the concentrations of vital substances—either in blood (the organism) or in the cytoplasm (individual cells) and mitochondria. Both systems (the organism and the cell) use specialized messenger molecules: the organism relies on hormones (for instance, insulin and glucagon), while the cell produces allosteric effectors (ATP, NADH, acetyl-CoA, malonyl-CoA, citrate, ADP, AMP, and NAD). Both systems also exploit specialized structural solutions which assist in regulation.

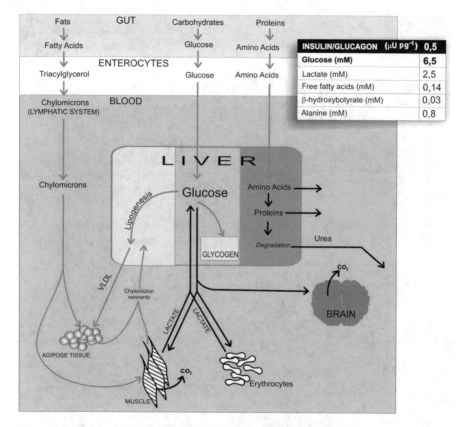

| INSULIN/GLUCAGON (μU pg⁻¹) | 0,5 |
|---|---|
| Glucose (mM) | 6,5 |
| Lactate (mM) | 2,5 |
| Free fatty acids (mM) | 0,14 |
| β-hydroxybotyrate (mM) | 0,03 |
| Alanine (mM) | 0,8 |

**Fig. 5.14** Collaborative energy management given an abundance of nutrients

## 5.5 Mutual Support Between Cells and the Organism: Interdependence Related to Gene Expression (Slow Effects)

If both the organism and its environment are in a stable state, the concentrations of enzymes and other proteins involved in biological functions are defined by biological programming. Under such conditions, interrelationship may be realized by means of standard activation or inhibition of messengers (depending on their exchange dynamics).

If, however, the environment exhibits high variability and adaptations need to be introduced, quantitative changes associated with synthesis of additional substances become important. One example is hypoxia caused by severe blood loss, where synthesis of new blood cells is greatly intensified. Although erythropoiesis is a continually occurring process, it can be significantly upregulated by the action of certain proteins triggered by hypoxic conditions. This process involves

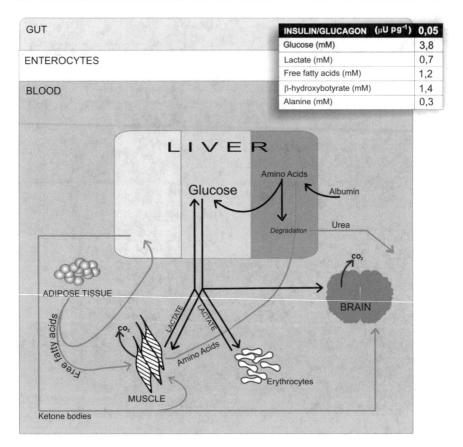

| INSULIN/GLUCAGON  (µU pg⁻¹) | 0,05 |
|---|---|
| Glucose (mM) | 3,8 |
| Lactate (mM) | 0,7 |
| Free fatty acids (mM) | 1,2 |
| β-hydroxybotyrate (mM) | 1,4 |
| Alanine (mM) | 0,3 |

**Fig. 5.15** Collaborative energy management under conditions of low nutrient availability (decreased blood glucose level)

differentiation of the erythrocyte line as well as actions related to hemoglobin synthesis (Fig. 5.17). Correct synthesis of hemoglobin, which consists of a protein unit and a porphyrin ring, depends on subprocesses which synthesize each of these components separately, as well as on the availability of iron. All subprocesses are subject to their own automatic control mechanisms, but they also exhibit cooperative behavior (Figs. 5.18 and 5.19). Since each occurs in a separate cellular compartment, they must all be coordinated in order to provide matching concentrations of various substances.

Excess heme halts the synthesis process while downregulating the globin synthesis inhibitor. This enables continued synthesis of hemoglobin, consuming the available heme.

In a separate process, the hemoglobin synthesis apparatus (mostly based on ribosomes) also undergoes rapid expansion.

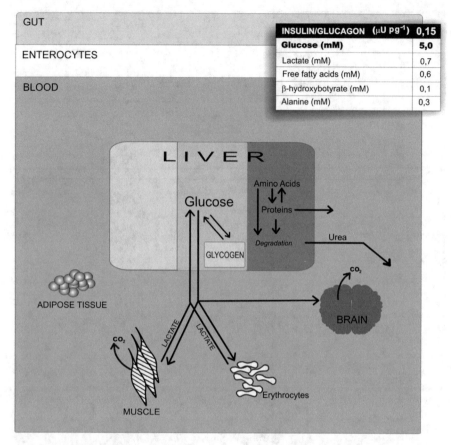

| INSULIN/GLUCAGON (µU pg⁻¹) | 0,15 |
|---|---|
| Glucose (mM) | 5,0 |
| Lactate (mM) | 0,7 |
| Free fatty acids (mM) | 0,6 |
| β-hydroxybotyrate (mM) | 0,1 |
| Alanine (mM) | 0,3 |

**Fig. 5.16** Collaborative energy management in hepatocytes and in blood when the activity of insulin and glucagon is moderate

Hormonal signals induce transcription of mRNA and rRNA in the nucleolus. Synthesis of ribosomal proteins occurs in the cytosol. Once synthesized, proteins migrate to the nucleolus where ribosomal subunits are formed and released back into the cytosol. At the same time, mRNA strands travel from the nucleus to the cytosol, where the required proteins can be synthesized (Fig. 5.17). The pathway for signals which trigger hemoglobin synthesis involves transcription and is controlled by hormones.

It should, however, be noted that translation is subject to its own hormonal control, which also assists in the synthesis of hemoglobin. A hemoglobin-synthesizing cell includes a special mechanism capable of inhibiting translation. Its action is dependent on the concentration of heme: deficiency of heme halts further synthesis of globin and therefore acts as a coordinator (Fig. 5.18).

Heme synthesis (occurring in the mitochondrion) is also controlled by hormones as it depends on the availability of iron, which has to be drawn from outside the cell.

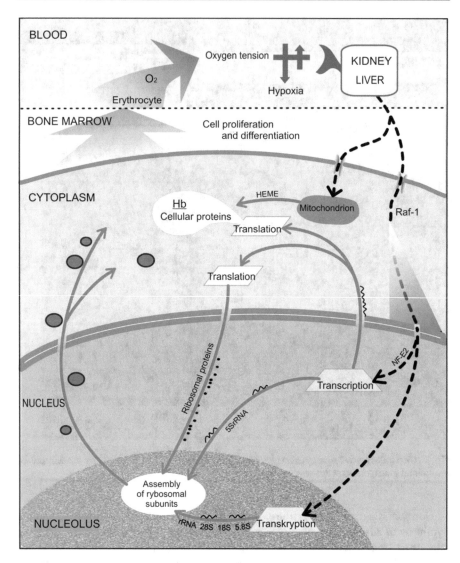

**Fig. 5.17** Regulation of the biological response to hypoxia. Cooperation and coordination of processes occurring in an erythroblast. Solid line indicates cooperation. Dashed line indicates coordination. Raf-1, kinase; NF-E2, transcription factor

Hemoglobin synthesis may proceed only if ample iron is available—thus, the response to hypoxia is conditioned by many independent (though mutually support-ive) control processes.

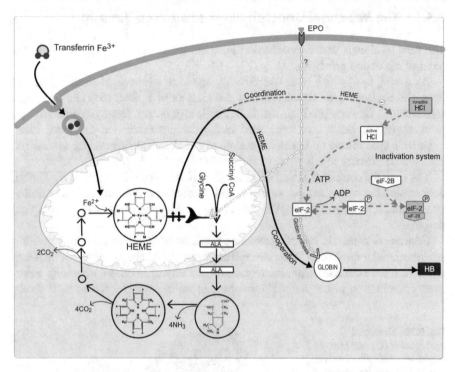

**Fig. 5.18** Cooperation and coordination in the synthesis of heme and globin. Right-hand side, globin synthesis control; HCI, heme-controlled inhibitor; eIF, eukaryotic initiation factor. Inactive forms grayed out

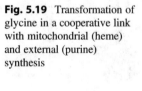

**Fig. 5.19** Transformation of glycine in a cooperative link with mitochondrial (heme) and external (purine) synthesis

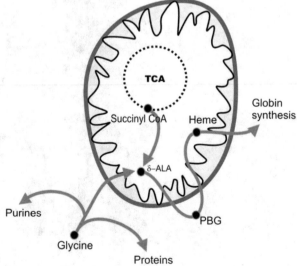

### 5.5.1    The Structural Underpinnings of Interrelationship

The flow of signals which coordinate biological processes can be modeled on the basis of regulation principles.

As a rule, the path of each coordinating signal is a branch of some regulatory loop. Once released, the signal seeks out the receptor of another loop and adjusts its sensitivity to its own product. Inside the cell, signals are usually represented by products of the cell's own controlled biological processes. On the other hand, organisms tend to rely on hormonal and neuronal signals, generated by specialized cells whose primary role is to assist in coordination (Fig. 5.20).

The need for correlated action becomes clear if we refer to the rules of regulation, determining the nature of both substrate and product of a given process (controlled by a feedback loop) and tracing the path of the product with respect to its biological purpose.

Coordination should be interpreted as a means of improving interrelationship whenever major deviations in the controlled quantity could be avoided (for instance, if a controlled substance exhibits undesirable activity or is otherwise toxic). However, coordination places an additional burden on biological systems as it forces

**Fig. 5.20**  Hormonal coordination of a complex response to a stimulus (such as stress). I, II, and III, hormonal cascade system levels; A, B, and C, effector loop cells

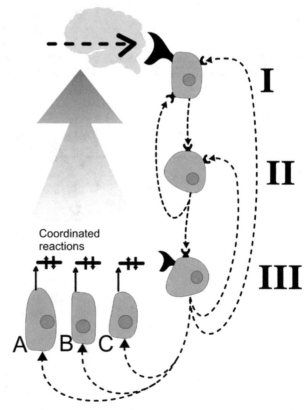

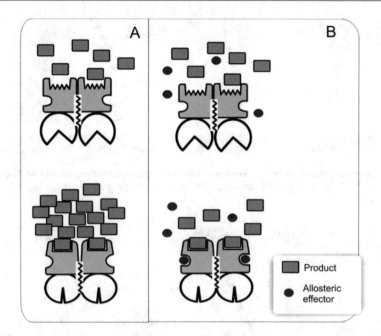

**Fig. 5.21** Model view of an internal cell receptor complex (enzymatic regulatory subunits) capable of binding its own product (rectangles) as well as the product of another coordinating system (black circles). (**a**) Conditions prior to interaction with the coordinating system (low and high product concentrations). (**b**) Conditions after binding the signal of the coordinating system (allosteric effector)

them to provide receptors for external signals (Fig. 5.21). A solution to this dilemma comes in the form of receptor proteins, capable of binding their own product in addition to the products of regulatory processes to which a given system is coupled. Such proteins are almost exclusively allosteric, consisting of subunits and able to bind effector molecules in specific sites—just like hemoglobin, which can bind BPG (2,3-bipohosphoglycerate) as well as protons by way of the Bohr effect.

The altered sensitivity of the receptor subunit is depicted as a change in the concentration of the controlled product, which, in turn, shuts down the controlling enzyme.

It appears that the need for coordination does not apply to all processes. Specifically, coordination can be avoided in processes which are of secondary importance to the cell. It is also unnecessary for sparse products, where significant fluctuations are not harmful to the cell or the organism.

Selected examples of metabolic pathways which require coordination are depicted in Fig. 5.22. As mentioned above, these are branching processes; moreover, their individual branches are often highly spontaneous and therefore irreversible. The figure presents processes (and parts thereof) which involve transmembrane transport (performed in separate organelles), where efficient transduction of products

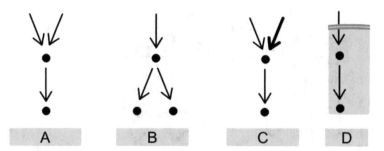

**Fig. 5.22** Examples of metabolic processes which require coordination. (**a** and **b**) Branching processes. (**c**) Variable inflow of substrates. (**d**) Process consisting of sub-stages separated by a membrane

**Fig. 5.23** Selective action of hormones upon "addressee" cells via specific receptors

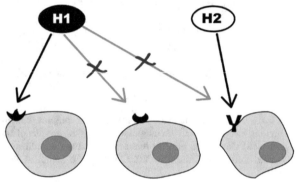

or substrates becomes difficult given significant variations in their availability or particularly rapid reaction rates.

Coordination of processes becomes easier if these processes share a common stage. The enzyme which catalyzes this stage (i.e., corresponds to a fork in the metabolic pathway) is naturally important for both intertwined processes. Such stages and the enzymes which catalyze them are called *key stages* and *key enzymes*, particularly if they affect vitally important processes. Examples include energy transfer mechanisms which involve synthesis of acetyl-CoA, glutamine, etc. These stages are natural targets for coordinating mechanisms.

A similar rule applies to hormonal coordination. A single hormone can control many processes by attaching itself to receptors exposed by many different types of cells (Fig. 5.23). Intracellular signal branching, resulting in the activation of interdependent processes, is also an example of complex coordination (Fig. 5.24). This type of action is observed, e.g., in the control of sequestration processes effected by insulin (Fig. 5.25). Each hormone, particularly a nonspecific one, can control a large number of processes simultaneously.

**Fig. 5.24** The impact of hormonal coordination on the cell

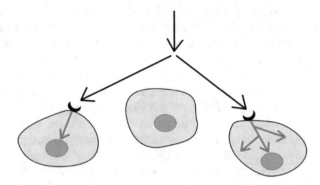

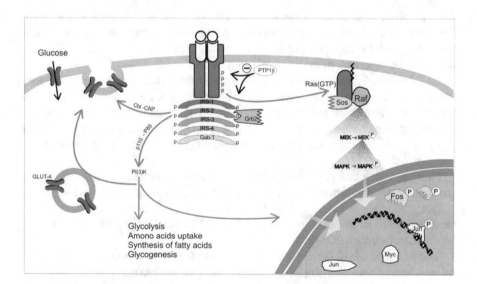

**Fig. 5.25** The examples of branched hormone (insulin) signaling path inside the cell

## 5.5.2   The Role of Common Metabolite in Complex Process Coordination

If a coordination pathway simultaneously affects many linked processes, one of its components may become an indicator for a whole set of biological mechanisms which together generate the required effect. This strategic selection of a common marker is an important "technical solution" in many integration processes. As an example, let us consider the coordination of nutrient sequestration which depends on blood glucose levels.

Glucose is always present in bloodstream where its concentration can be readily measured. It is also an integrating component in the management of lipid deposits as a source of oxaloacetate required for degradation of acetyl-CoA. By penetrating adipose cells, glucose becomes a source of glycerol-3-phosphate, powering fatty acid sequestration (lipogenesis). As glucose processing is coupled to many energy management mechanisms, each such mechanism may affect blood glucose levels.

The storage of glucose in the form of glycogen is only meant as a buffer (no more than ~600 kcal is available at any given time, compared to the average basal metabolic rate of approximately 1600 kcal for a typical human). Thus, glucose sequestration and subsequent release must be a highly dynamic process. High concentrations of glucose affect the spontaneity of its intracellular metabolic pathways. Glucose is a good coordinator of energy management processes given its ubiquity, universality, and the critical role it plays in many tissues such as muscles, blood, and brain matter. It should also be noted that glucose is effectively the only nutrient whose concentration in blood is never subject to major fluctuations—thus, any changes in its levels are indicative of the state of energy management processes in the organism as a whole.

Glucose directly affects the release of two basic hormones: insulin and glucagon. Other hormones such as catecholamines or adrenal cortex hormones are less important for managing the organism's energy stores. They usually come into play under stress conditions (catecholamines) or during prolonged periods of starvation when glycogen stores are depleted and the organism must draw energy from amino acids (adrenal cortex hormones).

Nutrient sequestration can be affected—via increased insulin release—by certain gastrointestinal hormones (incretins) such as GIP (*gastric inhibitory polypeptide*), whose production is stimulated by the presence of glucose and fatty acids in the small intestine: an interesting example of how the organism predicts and prepares itself for assimilation of nutrients.

The presence of substances which act as universal control indicators is biologically advantageous and can be observed in many different mechanisms. For instance, the entire nitrogen cycle in the *E. coli* bacteria is based on measuring the concentration of glycine and alanine. Bacterial glutamine synthetase consists of two subunits and has been shown to contain no less than nine allosteric binding sites upon which coordinating signals may act. These signals are, in turn, generated by cooperating processes and include carbamoyl phosphate, glucosamine-6-phosphate, AMP, CTP, and several amino acids (tryptophan, histidine, serine, glycine, and alanine). It should be noted that not all amino acids are actively involved in controlling the activity of the synthetase complex: from among the available molecules, alanine and glycine have been selected as the most common and the most intimately tied to the bacterial nitrogen cycle. A similar function appears to be performed by cyclins, whose synthesis is coupled (both in terms of reactions and substrates) to cell division, making them a good indicator of the division process.

### 5.5.3  Signal Effectiveness and the Structuring of Mutual Relations in Metabolism

In order to better understand the efficiency of regulation, cooperation, and coordination, we should consider the effectiveness of biological signals. This property is determined by the intensity of the signal and the sensitivity of its receptor; however it may also relate to the signal's influence (the scope of activation or inhibition effects it produces).

Comparing the relative effects of various allosteric effectors on coordinated processes enables us to define a clear structure of coordination. The gradation of signal effectiveness is subject to certain rules. Hormones are usually far more potent than concentration-mediated signals (which characterize intracellular coordination) as their action involves signal amplification and covalent modifications of receptors. It is, however, more difficult to ascertain the relative hierarchy of the cell's own signaling pathways. This hierarchy depends on the sensitivity of the regulatory enzyme to a given signal (which is constant) and on the concentration of signal molecules (which may vary).

For instance, we should expect that the effect of ATP—an allosteric effector involved in the cell's energy management processes—upon mitochondrial enzymes will be stronger than its corresponding influence on cytosolic enzymes as ATP is highly diluted in cytosol compared to its mitochondrial concentration.

Glycolysis and the Krebs cycle involve several checkpoints where they may be inhibited by ATP. Proper description of these coordination processes requires a comparative assessment of ATP-mediated inhibition at each checkpoint for a given concentration of ATP.

In order to ascertain the specificity of coordination, we must first determine the affinity of the enzyme regulatory subunit (or other receptor systems) to the allosteric effector as well as the sensitivity of the entire system to the coordinating signal.

Variable signal affinity can be observed in many circumstances. One example is the coordination of ribosomal protein and rRNA synthesis, based on the ability of proteins to form complexes with rRNA and mRNA with differing levels of affinity.

Once synthesized, ribosomal proteins preferentially bind to rRNA (whose synthesis is mediated by hormones), creating ribosomes. This process continues until all available rRNA is consumed and free ribosomal proteins begin to build up. At this point freshly synthesized proteins start binding to the promoter fragment of their own RNA template, effectively halting the synthesis process (Fig. 5.26).

A similar problem occurs in coordination of cell division processes, where specialized proteins called cyclins are synthesized in tandem with division mediators. The energy and material expenses incurred by cyclin synthesis hamper the synthesis of proteins involved in cell division; however the latter group must be prioritized. Thus, cyclins amass at a slower rate, reaching their peak concentration only after the synthesis of the required division proteins and nucleic acids has concluded. Rapid degradation of cyclins triggers the next step in the division cascade. In this way a time- and energy-consuming process may act as a controller for other mechanisms associated with cell division. Instead of a "biological clock,"

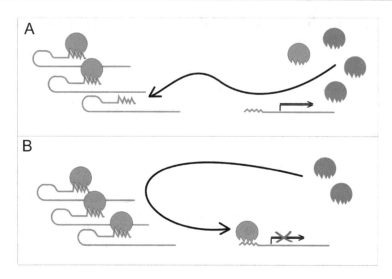

**Fig. 5.26** Coordination of the synthesis of ribosomal proteins in the ribosome construction process. Ribosomal proteins exhibit higher affinity to rRNA and lower affinity to their own mRNA. Once rRNA is depleted, they begin binding to mRNA, halting their own synthesis

we are dealing with coordinating action, dependent—much like the cell itself—on the available energy stores and supplies of substrates.

### 5.5.4  Interrelationship in Times of Crisis: Safety Valves

Automatic regulation and coordination mechanisms allow cells to maintain proper substance concentrations and activity levels in spite of environmental changes. However, adaptability has its limits. A particularly acute crisis emerges when the cellular metabolic pathways are blocked by insufficient availability of oxygen. Energy requirements cannot be met if the cell is unable to metabolize nutrients. This is evidenced, e.g., in skeletal muscle cells, which on the one hand require an extensive network of blood vessels but on the other hand must be able to contract, limiting their ability to absorb oxygen from the bloodstream. One consequence of this apparent paradox is the synthesis of lactic acid. Given an insufficient supply of oxygen, this byproduct cannot be further metabolized and therefore builds up in the muscle cell. High levels of lactic acid threaten the cell by lowering its pH, increasing osmotic pressure, and—most importantly—inhibiting glycolysis by altering its oxidation-reduction potential and interfering with enzymes which participate in breaking down glucose.

If the cellular regulatory mechanisms are unable to prevent excessive buildup of lactic acid, the only solution is to expel this unwanted byproduct into the blood-stream where acidification can be counteracted by other regulatory systems (such as the liver, capable of neutralizing lactic acid). This mechanism can be compared to a safety valve which prevents the destruction of an overpressurized boiler by venting excess steam. A similar situation occurs during starvation when the relative deficiency of oxaloacetic acid inhibits the Krebs cycle and hampers further metabolism of acetyl-CoA (extracted from fatty acid degradation), resulting in increased production of ketone bodies which must also be ejected into the bloodstream. Oxaloacetate acid synthesis is largely dependent on the supply of glucose (a key source of pyruvate).

Many organisms have evolved coordination mechanisms which enable them to handle the byproducts of metabolism appearing in the bloodstream. Naturally, single-cell organisms have no such problems and simply expel unneeded substances (alcohol, lactate, acetoacetate, etc.) until their concentration in the environment halts cell proliferation.

Lactic acid is produced mainly by muscle cells and erythrocytes and metabolized by hepatocytes. A similar mechanism applies to ketone bodies, which are generated in the liver and broken down in the brain, heart, kidneys, and skeletal muscles.

## 5.6    Specialization of Cell Interrelationship

Specialization results in functional differentiation and requires active relationship if the differentiated tissues are to constitute a biological entity (a cell or an organism). Examples of specialization include organelles (compartments) in the cell and organs in the organism, although each of these groups follows different principles. Cell organelles are self-contained intracellular subunits which assist the cell in achieving its strategic goals. They facilitate cooperation and coordination but also enable separation of processes which would otherwise interfere with one another (such as synthesis and degradation). An example of this mechanism is β-oxidation and synthesis of fatty acids. Furthermore, some processes are potentially harmful to their environment and need to be isolated (this includes lysosome activity).

Given our present state of knowledge, it would be difficult to accurately and unambiguously describe the strategies of intracellular compartmentalization. Some enzymes change their location as a result of evolutionary processes. For instance, rhodanese has migrated from the cytoplasm to the mitochondrion in the course of evolution. Explaining this shift would require good knowledge of the interactions between rhodanese and the intracellular environment; however it also proves that at least in some cases enzyme location is not crucially important.

In light of the above facts, we may ask why glycolysis—a degradation process—is not itself restricted to the mitochondrion. The answer lies in the natural strategies applied to separation and aggregation of intracellular processes. Assuming that separation of contradictory processes, while useful, is not of key importance enables us to consider other factors which may influence their location. In the presented case, synthesis of ATP and other energy carriers is more important than any potential benefits derived from physical separation of the relevant metabolic pathways (note that mitochondrion-independent glycolysis is an important source of ATP).

A crucial property of glycolytic ATP synthesis is its independence of an external oxygen supply. In contrast, the Krebs cycle and conjugated with it other mitochondrial processes slow down whenever oxygen becomes scarce. Under such circumstances an oxygen-free energy supply becomes critically important and cannot be subordinated to mitochondrial processes. Of course, relinquishing physical separation of glycolysis components introduces the need for accurate control and coordination, which in turn explains the importance of regulatory processes associated with glucose metabolism.

Considering the organism as a whole, specialization should be understood as a consequence of cell differentiation. On this level interrelation means the difference between a pool of random cells and an efficient organ which can assist its host organism in maintaining homeostasis. In animals cooperation is enabled by transporting substances via the bloodstream, while the coordination of differentiated tissues occurs by way of hormones or (in certain cases) neuronal signaling.

The function of the organism is based on a homeostasis program "encoded" in the structure of its receptors.

In general, the term *organism* is applied to any group of independent, differentiated units which together constitute a higher-level structure and coordinate their actions in pursuit of a common goal. This process opens up avenues of development not available to undifferentiated units such as cells.

The word *organism* may also be used in relation to the social structure of a state or an insect colony comprising individual castes which communicate through chemical means (pheromones) and through physical contact. Examples of this phenomenon include ant and bee colonies.

In humans, the "colony" model can best be observed in the immune system, which also consists of specialized "castes" of cells—workers (B and T lymphocytes) and soldiers (monocytes, killer cells, and phages). Both groups originate from similar nonspecific proliferation centers (hematopoietic stem cells, which can be compared to ant colony queens), and both exhibit the ability to gain a specific "personality" via interaction with the histocompatibility complex (much like pheromones, lymphokines, or direct physical contact among social insects). Moreover, immune system cells are also able to acquire experience.

The level of similarities between such disparate biological systems suggests a common organizational strategy based on the principles of information theory.

## 5.7   Microbiome

A peculiar mode of co-action is encountered in human and animal organisms which form mutualistic relationships with bacteria present in their organs—especially in the gut. The number of bacteria present in a typical organism is typically far greater than the number of cells which make up the organism itself. Clearly, the presence of such a large quantity of microbiota must have an effect on the organism. Given that—under normal circumstances—these microbiota do not cause disease and in many cases actively promote the organism's well-being, their activity should be regarded as beneficial, even if they are not integrally linked to the organism. This so-called microbiome is highly diverse, and its composition strongly depends on the host's dietary habits, which is why it varies from organism to organism. The microbiome's role is not limited to digestion. While it clearly influences metabolism, it is equally important as a source of stimuli for the immune system, exploiting the activity of B cells migrating from germinal centers. As a result, this bacterial system enjoys a mutualistic relationship with its host.

## 5.8   Bacterial Bioengineering Techniques

Explaining the structure of the genetic code and the basic principles involved in its construction paved the way toward possible interventional procedures, roughly referred to as "genetic engineering." The necessary prerequisite of such interventions is the ability to accurately cleave the DNA helix in specific places, in order to introduce the desired modifications and mutations, and also to copy specific fragments of the genetic material.

As can be expected, genetic engineering techniques reflect mechanisms which have long been employed by prokaryotes. Bacteria in particular have evolved intracellular defense mechanism against their principal foe: phages, i.e., entities which penetrate bacterial cells, injecting their own DNA. Such mechanisms work by identifying and destroying alien DNA.

Since bacteria cannot rely on external defensive structures (unlike higher organisms), they had to devise intracellular mechanisms capable of detection and selective hydrolysis of alien DNA. This mechanism has been initially repurposed in human genetic engineering. Its specificity is achieved by accurate action of bacterial enzymes known as restriction enzymes, capable of recognizing palindromic sequences. Such sequences are characterized by repetition and symmetry, as illustrated Fig. 5.27.

Restriction enzymes are capable of recognizing specific sequences and work by cleaving DNA strands in two slightly offset places, leaving behind an uneven tip known as a sticky end. These ends are indeed "sticky" given that they may spontaneously, noncovalently reassociate, restoring the original connection, which is subsequently covalently reinforced by ligase (Fig. 5.28).

Palindromic sequences are ubiquitous and can be found in DNA derived from various organisms. By cleaving its affine palindrome, the restriction enzyme fragments DNA. The resulting fragments may reassociate while incorporating other external DNA (created in a separate cleaving process, mediated by the same enzyme), since all its ends remain "sticky"—which is due to the fact that they share a complementary sequence.

This mechanism, discovered in the bacterial domain, has been applied to replicate genetic material and to synthesize specific proteins.

The bacterial genome is arranged in a large circular chromosome and many smaller, chromosome-like circular structures called plasmids, some of which are capable of replication. Genetic engineering utilizes entities known as vectors—small

**Fig. 5.27** Schematic view of using restriction enzymes to cleave DNA in specific places, leaving behind so-called sticky ends, as well as employing bacterial plasmids as carriers of external DNA, pasted in the process of replicating genetic material

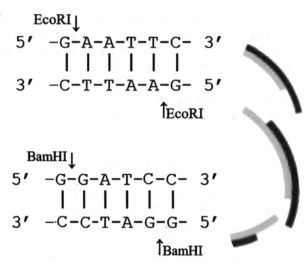

plasmids which have been modified to contain a single palindromic site corresponding to a specific restrictase. By "opening up" the circular plasmid, the restriction enzyme primes it for attachment of external DNA (cleaved by the same restrictase and therefore possessing matching sticky ends). The modified plasmid, containing alien genetic information, is then reintroduced into the bacterial cell and replicated. This "cuckoo's egg" strategy enables us to produce a large quantity of specific genetic materials, which is inserted into the plasmid and subsequently isolated using the same restrictase as before (Fig. 5.28).

Another more advanced technique which bacteria employ to combat phages and which can also be utilized by genetic engineering is referred to as CRISPR (clustered regularly interspaced short palindromic repeats). It assumes the form of a genetically encoded matrix for a RNA sequence which directs the attached CAS nuclease to cleave both strands of the phage's DNA at a specific point. The operon involved in this process represents a unique intracellular bacterial defensive measure, capable of remembering specific features of its attack targets, including the phage's DNA. The associated information packets, encoded in the bacterial chromosome, carry information of alien (phage) DNA with which the bacteria have previously come in contact. Such information is interspersed with short palindromic sequences. The resulting RNA can recognize the phage's DNA, since its own DNA matrix is, in fact, the phage's DNA which has been fragmented, integrated with the bacterial genome (bracketed by palindromic fragments) and subject to inheritance as a form of long-term database for the bacterial anti-phage defense system. In its active form, the system assumes the form of RNA associated with CAS nuclease (in the case of the commonly analyzed *Streptococcus pyogenes*—CAS[9]), which becomes active by binding to RNA. CAS nucleases consist of two domains referred to as HNH and RuvC. Each cleaves a specific strand of the phage's DNA, with the HNH domain

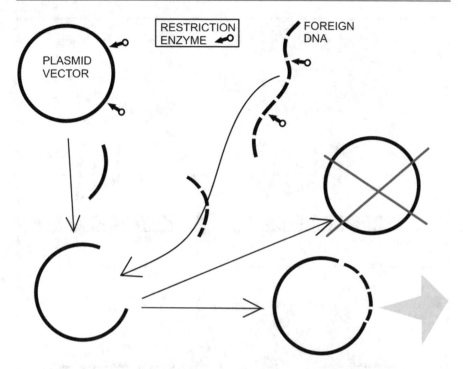

**Fig. 5.28** Mechanism of DNA amplification with the aid of bacterial plasmids

targeting the strand directly referenced by the bacterial RNA (in complex with CRISPR).

Recognition of alien DNA is not, however, based solely on interaction of a specific RNA strand with the target DNA. An additional condition involves interaction between nuclease and a short (two to four nucleotides) fragment of the target DNA, referred to as the protospacer adjacent motif (PAM). PAM is directly proximate to the fragment which corresponds to the RNA template, and its interaction with nuclease initiates splitting of the target DNA strands, enabling RNA to form a complex with one of them.

The structure and activity of the CRISPRCAS system both confirm the ubiquitous mechanism whereby RNA is used as a "pointer," leading active proteins to their intended targets, whether DNA- or RNA-based (Fig. 5.29).

Eukaryotic cells employ a similar system, centered upon the so-called Argonaute protein, which exhibits nuclease activity. This system has recently attracted increased attention as a possible alternative for CRISPRCAS, endowed by greater specificity and smaller size, both of which facilitate integration with carrier viruses.

The CRISPRCAS system has proven highly versatile owing to its capacity for easy modification and the resulting wide range of applications in biology and medicine. Modifications may involve the RNA itself as well as the interacting proteins. Some authors have reported successful attachment (via fusion) of further

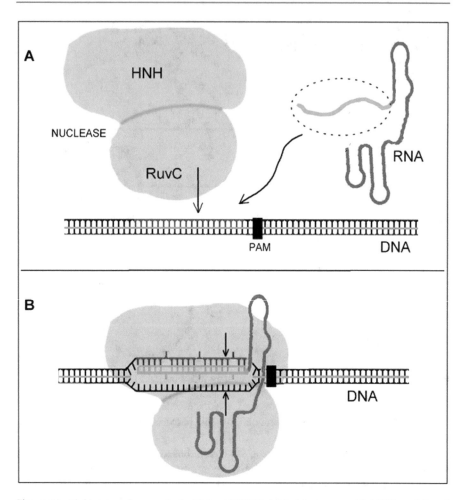

**Fig. 5.29** Linking together products of the CRISPR bacterial operon with HNH and RuvC domains of the CAS nuclease in an attack on a recognized DNA sequence

proteins, which provide additional activity. CRISPR has so far found use as an extremely sensitive detector of viral infections, revealing the presence of viruses in the organism long before any clinical symptoms emerge. Nevertheless, its in vivo use for introducing targeted genomic modifications encounters obstacles related to false positives and off-target mutations. The technique is still being perfected, and its great promise suggests that CRISPR may, in the near future, become the mainstay of clinically relevant genetic engineering techniques.

In vivo techniques also employ AAVs (adeno-associated viruses) as carrier vectors.

Another powerful technique allowing amplification of DNA synthesis using selected DNA chain fragments as the template is known as polymerase chain reaction (PCR). It was devised in 1984 by Kary Mullis. The technique derived from thermophilic bacteria heat stable polymerase allowing polymerization of DNA after having its chains separated simply by temperature (95 °C, 5 s) applies. The rapidly lowered temperature to 54 °C allows then the attachment of primers, and when temperature is again elevated (72 °C, temperature optimal for polymerase activity), the process of DNA synthesis starts. This procedure may be repeated again and again producing the increasing amount of selected DNA fragments. Thus finally in the presence of substrates (active nucleotides), primers kept in the access, and heat stable polymerase, only the suitable manipulation of temperature is necessary to run the DNA synthesis.

## 5.9 Hypothesis

### 5.9.1 Carcinogenesis

The mechanism of neoplastic transformation remains unclear despite numerous efforts to understand it. In particular, an integrated approach to the problem from a molecular perspective is yet to be devised.

It seems clear that the process is driven by mutations; however, mutability is a very broad concept, and its links to neoplastic transformation are not understood in detail. In most cases a cell which has undergone such transformation is found to contain numerous mutations, causing significant damage to its genome. It would be natural to expect such cells to exhibit poor viability—which runs counter to the observation that cancer cells are often highly active and prolific. What is more, these active, rapidly dividing cells do not resemble ordinary building blocks of the organism—they do not conform to restrictions imposed by the organism and should instead be treated as alien, harmful agents.

Uncontrolled division and loss of intended biological function are the hallmarks of genetic reprogramming which causes otherwise ordinary cells to break free from the rules to which all healthy cells must conform. The organism is a system composed of varied but cooperating cells, all of which carry out a common program, ensuring homeostasis and enabling the organism to function as a coherent whole. This situation may be compared to a political system, involving a state and its citizens—distinct but collaborating entities. The latter are subordinate to the former, while the former ensures coordinated action.

In both cases—i.e., with regard to cells as well as citizens—subordination relies on signals which are mandatory in character. Thus, both systems have a hierarchical nature. In the case of the state, this hierarchy is enforced by its administration, whereas in an organism, it assumes the form of signals, which—unlike intracellular signals—depends on covalent bonding (independent of substrate concentration) and very strong amplification, typically facilitated by cascading processes.

Given such conditions, it becomes imperative to prevent the command signal from persisting beyond its intended timeframe. As a result, cells have evolved mechanisms which enable such signals to be attenuated or suppressed entirely.

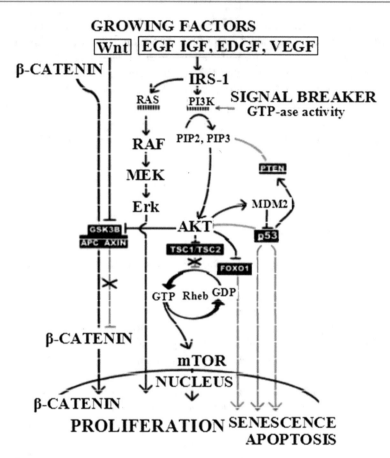

**Fig. 5.30** Command signals guiding proliferation, with their associated suppression components (in rectangular frames)

This function falls to the so-called suppression systems associated with signaling pathways (Fig. 5.30). Their action usually manifests itself as a GTPase-dependent process—as in the case of protein G, for which GTP activates signals, while its hydrolysis attenuates them or, alternatively, the action of phosphatases which suppress phosphorylation-related activity and determine the activity of kinases, such as tyrosine kinase. This suppressive effect may be present at all times and become amplified along with the amplification of command signals. Examples of such systems include suppressors involved in cell proliferation, such as p53, APC, axin, and others (Fig. 5.31).

The role of suppressor systems is critical in ensuring balanced and controlled command signal activity. Mutations which inactivate suppressor genes in the cell division process lead to uncontrolled proliferation and are implicated in neoplastic transformation.

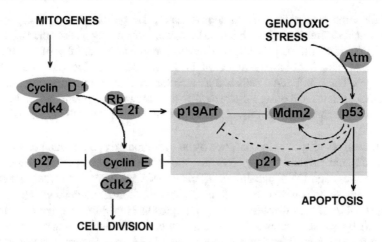

**Fig. 5.31** p53 and its basic cellular interactions, revealing the effect of mutations. The figure illustrates the permanent standby activity of the p53 suppressor system in the scope of cell proliferation

In the course of evolution, independent cells became involved in higher-order structures—organisms—and had to adapt their "programming" accordingly. In particular, cells which form part of an organism had to:

1. Undergo differentiation, subject to the principles of epigenetics.
2. Become susceptible to controlled senescence and apoptosis to enable selective development and supersedure of cells which have accumulated too many genetic defects during their lifetime. This goal is achieved by progressive downregulation of telomerase, which imposes an effective limit on the number of divisions a cell can undergo, or by mechanical elimination of cells from the organism—e.g., by exfoliation of epithelial cells,
3. Limit their proliferative activity in accordance with "stop" signals received from neighboring cells, in order to ensure formation of properly structured organs. This mechanism is also known as contact inhibition.
4. Optimize their respiratory efficiency.

The emergence of organisms could be regarded as a major evolutionary breakthrough—as well as a breakthrough in terms of cellular programming, where individual cells had to forgo some of their autonomy in order to obey commands issued by the organism.

Neoplastic cells break free from this command hierarchy while retaining their viability and ability to proliferate (despite being saddled with numerous mutations). This phenomenon could be compared to an act of rebellion against the organism. Such selective outcome of mutations may be explained by referring to the process of evolution and to the specific nature of certain mechanisms which emerged in its course.

While eukaryotic cells evolved approximately 1.6 billion years ago, organisms have a much shorter history, with the earliest such biota dating back to the Ediacaran period (ca. 650 million years ago). Complex organisms—initially of the marine variety and later land-based—appeared in the Cambrian period (ca. 100 million years ago). Thus, individual cells had a much longer time to fine-tune their internal regulatory systems and ensure their stability than organisms, which rely on broader regulatory mechanisms to ensure coordination among various types of cells (Fig. 5.32).

Sufficient stability may be conferred by an evolutionarily conditioned set of genes which do not mutate easily due to the presence of proteinic safeguards (nucleosomes) and DNA packing or efficient DNA repair mechanisms. The same effect may also be achieved by systems capable of eliminating cells which contain damaged DNA or are otherwise aberrant (apoptosis). However, a similar outcome is produced by suitable suppression of command signals, which—among others— prevent excessive proliferation and thus protect DNA from incurring unacceptable mutational pressure. Nature provides us with many mechanisms which may potentially mitigate neoplastic pathologies. Examples include anti-cancer compounds which influence regulatory mechanisms related to cell proliferation, such as thalidomide or rapamycin. Given the availability of such mechanisms, an interesting question is why neoplastic transformation occurs at all, and why it poses such a serious threat to organisms. The simplest answer is that there has not been enough time (in evolutionary terms) for organisms to evolve foolproof protective mechanisms. Another point which can be made is that it is often counterproductive (again, in terms of evolutionary gains) to prolong the lifespan of individual organisms—thus, no corresponding mechanisms have evolved over the long term.

DNA mutation is a progressive process, spurred by random errors which occur naturally during the replication cycle. A serious threat emerges when the rate of mutations is accelerated—whether by external or internal factors. The former include ionizing radiation, aggressive chemical compounds, or viral infections, while the latter mainly involve free radicals (a natural byproduct of the respiratory cycle) or hormonal deviations. DNA's susceptibility to mutagens increases when immediate repair is hampered or impossible. This often occurs when a single DNA strand is exposed for a long time in the process of replication or transcription. A rapidly dividing cell is particularly prone to accumulating mutations—especially, when there is not enough time to properly pack its DNA and repair emerging errors. Damage to genes which participate in suppressing command signals in division control pathways is especially dangerous. Breakdown of proliferative control— which is the main mechanism by which the organism exerts control over individual cells—causes the cell to become independent of the organism and switch over to its evolutionarily conditioned survival program, with no regard for the needs of the organism as a whole. This phenomenon can manifest itself through (1) ignoring command signals, particularly contact inhibition effects; (2) activation of telomerase, which effectively renders the cell immortal; (3) regression of differentiation through remodeling of chromatin; and (4) preference for fermentation over aerobic respiration (the so-called Warburg effect).

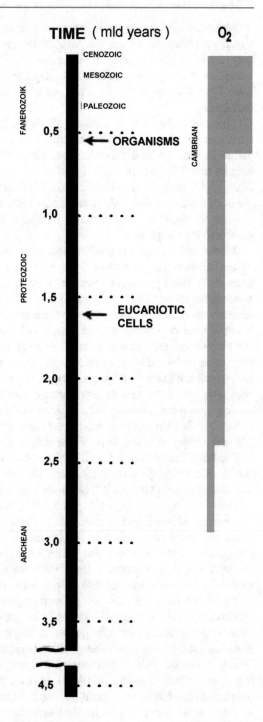

**Fig. 5.32** Timeline of the evolution of organisms against the backdrop of geological periods

Mutations which emerge in the course of replication preferentially affect those signaling pathways which are evolutionarily younger and devoid of suitable safeguards—such as inherent structural stability and resilience of attenuating components of signaling mechanisms, capable of counteracting excessive proliferation. Disabling such mechanisms renders the cell susceptible to uncontrolled divisions.

The need for and operation of signal attenuation mechanisms may be schematically demonstrated by referring to the emergence of statehood in societies and to its subsequent democratization. Here, analogues of attenuation mechanisms may be found in institutions designed to counteract uncontrolled exercise of executive power, such as the Constitutional Tribunal, the Ombudsman's Office, the National Judiciary Council, some media outlets, etc. Creation and fine-tuning of systems which act in opposition to top-down command pathways and can attenuate them require a long time.

Mutational changes in signaling pathways are more commonly observed in cases of cancer than pure statistics (e.g., Gaussian distribution) would suggest. One example is the Tp53 gene, found to be damaged in over 50% of cancers. Such nonrandom susceptibility to damage in mechanisms which control command and suppression activity is also evident in the irreversibility of neoplastic transformation. Proof of selective susceptibility to mutational pressure is found in the different rates at which tumors are found in various types of organisms and individual organs. Here, edge cases include laboratory mice (which are highly susceptible) and African mole rats, which can live over 30 years and almost never develop tumors. In the latter case, high resilience is thought to result from particularly efficient protein synthesis and folding processes, believed to be assisted by RNA S28-mediated ribosome activity and synthesis of macromolecular hyaluronic acid, associated with upregulation of p16 and p27 genes, which improve the efficiency of contact inhibition. This confirms that mechanisms reducing the risk of neoplastic transformation may emerge in the course of evolution. As organisms age, the risk of carcinogenesis increases due to accumulation of mutations but also—likely—due to reduced efficiency of protein synthesis and degradation mechanisms, all of which are vitally important.

Thus, neoplastic transformation can be described as a process through which cells break free from their link with the organism and revert to existing as individual, independent entities while also forfeiting their differentiation (partly or fully). According to this hypothesis, tumor cells "reenact" an evolutionarily ancient program which existed prior to the emergence of complex organisms.

From the point of view of pathophysiology, the direct cause of neoplastic transformation is usually associated with excessive, unchecked cell divisions, increasing disorder in the arrangement of chromatin and the resulting increases in mutation rate. Interestingly, the general characteristic of transformed cells tends to be similar despite differences in their initial epigenetic configuration. When comparing genes which undergo mutations in various types of cells, leading to their neoplastic transformation, it turns out that their functional properties are also similar. In short, these genes can be divided into three groups: (1) genes associated with signaling pathways (mainly suppressors); (2) genes associated with maintaining

|   | I | II | III |
|---|---|---|---|
| A | TET2 <br> DNMT3A <br> IDH1 <br> IDH2 <br> EZH2 | ARIDIA | ARIDIA <br> ARID <br> MLL2.3 <br> KDM6A <br> CREBBP <br> EP300 |
| B | RAD21 <br> RUNX1 <br> STAG2 <br> SMC3 <br> WT1 <br> CEBPA | CDC27 <br> FBXW7 | BRCA2 <br> MALATI <br> ERCC2 <br> CDKN1A(p21) |
| C | KRAS <br> FLT3 <br> NPM1 <br> PTPNII <br> KIT | TP53 <br> PIK3CA <br> KRAS <br> APC <br> AXIN2 <br> SMAD2.4 <br> TCF7L2 <br> BRAF <br> ACVR1B | TP53 <br> ERBB3 <br> EGFR3 <br> TBCID12 |

**Fig. 5.33** Groups of genes with different origin but similar functional profiles, all participating in enforcing cell-organism integration and often implicated in neoplastic transformation (as mutants): I, ureters; II, large intestine; III, leukemia

proper DNA structure and DNA repair, synthesis, and packing; and (3) genes directly associated with epigenetic processes and chromatin structure (Fig. 5.33).

Mutations in this set of genes often result in accelerated cell division since they mainly affect signal suppressors. Increased mutability is a natural consequence, and in this process the evolutionarily younger genetic programming—which is not equipped with strong protective mechanisms—is often disrupted, leaving more ancient processes intact.

Maintaining homeostasis in the scope of signaling pathways calls for a set of interlinked suppression systems. The resulting network is quite complex and appears more susceptible to disruption than the command system itself. Notably, certain gene families are strongly implicated in mutation-induced carcinogenesis—these include Tp53, WHL, FLT3, APC, ARIDIA, PIK3CA, NPM1, KRAS, FBXW7, MLL, and KDM6A. Tp53 is particularly noteworthy as it is closely tied to apoptosis and to suppression of telomerase. It therefore seems likely that the main function of genes which often undergo mutations in cases of cancer is to safeguard the link between the individual cell and the organism to which that cell belongs. Inactivation of such genes opens the door to neoplastic transformation.

## 5.9.2 The Criteria of Life

Expanding biological knowledge allows us to study the strategies used by nature. However, we still lack a satisfactory definition of the very notion of life. Self-organized, autonomous biological systems are usually said to be "alive"; however this inclusive criterion does not posit any specific boundaries. The question whether certain structures (particularly subcellular ones) are alive remains unresolved. Moreover, there is still no scientific consensus regarding the procedural definition of life.

As can be expected, this issue has been approached by many great researchers over the course of history. It seems that the definition which most closely matches our modern scientific knowledge is the one proposed by Claude Bernard, who focused on the independence of biological systems, i.e., their ability to function under varying environmental conditions. Increasing freedom of action—corresponding to increases in the complexity of biological systems—appears to be the most general defining characteristic of life. Freedom of action can therefore be treated as a measure of the complexity of an automaton, reflected by its capability to make autonomous decisions.

The properties and characteristics of structures which we call "alive" may be explained by their automatic behavior. It would, however, be misleading to fully equate a biological system to an automaton. Modern technology, particularly robotics, creates automata which sometimes closely resemble living organisms, blurring the border between life and technology. Nevertheless, biological entities differ from even the most intricate robots: they are governed by specific rules which determine the structural and functional differences between various types of organisms. Life is subject to certain restrictions not observed in robotics. For instance, a fundamental aspect of nature is the phenomenon of programmed death—the result of biological strategies which focus on the survival of the entire systems rather than individual entities, as a means of ensuring harmony and maintaining equilibrium. All properly constructed biological entities must include death as part of their natural programming. In cells, this programming is evidenced by limitations in the possible number of divisions, along with senescence and apoptosis. The inclusion of a terminating function (death) in a biological program calls for a clock-like mechanism, capable of measuring time. Cyclical processes, such as the circadian rhythm, which fulfills the role of a "biological clock," are crucial for living organisms. Entities whose programming does not involve the possibility of death—for instance, tumor cells—are considered pathological. (Note, however, that this group does not include gametes, where only the genetic information can be called "immortal"—individual gametes can and do undergo controlled death.) Cells which do not age or die are inconsistent with the principles of nature. In this aspect, the definition of life must necessarily be enumerative: nature itself determines what is and what isn't alive. Close study of biological entities straddling the border between the animate and inanimate worlds reveals the importance of controlled death in natural systems.

By analyzing and comparing the genomes of primitive organisms and other biological entities, we can quickly derive a minimal set of genes and functions required to support what is commonly understood as life. Let us focus on the

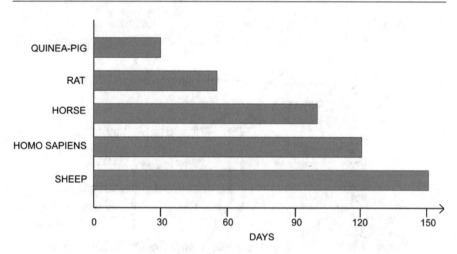

**Fig. 5.34** Average duration of life of erythrocytes in various animal species

erythrocyte, which is clearly a biological structure (even though it cannot reproduce). The fact that it derives from a fully functional cell makes it a good test subject.
Erythrocytes retain automatic control of metabolic processes (steady state),
maintaining proper ion gradients and oxidation-reduction potentials which enable
them to perform their function. They also actively counteract deviations which
emerge as a result of cyclical migrations from the lungs to other tissues. What is
more, they are subject to rapid degradation in a process which clearly defines the end
of their usefulness (Fig. 5.34). Maintaining a steady internal environment despite
drastic external changes is an important characteristic of healthy erythrocytes. In a
human organism, each erythrocyte visits the lung capillaries approximately once per
minute, releasing carbon dioxide and protons (this is called the Haldane effect). The
oxygen-rich environment of the lung increases the odds of encountering dangerous
reactive compounds (oxidative stress). By the same token, traveling through the
acidic environment of metabolically active tissues alters the production and degradation rates of various intermediate substances in key metabolic pathways. Of note is
also the relative difference in temperatures—from 28 °C in lung tissue to 40 °C in
certain areas of the liver. In the kidneys, erythrocytes are subject to a rapid increase
in osmolarity. Finally, the width of a typical erythrocyte is larger than the diameter of
a capillary which means that traveling through narrow capillaries causes friction and
distorts the cellular cytoskeleton. Counteracting all these changes requires the cell to
expend energy (in the form of ATP and its derivatives reduced by NAD and NADP).

ATP synthesis in human erythrocytes depends on glycolysis (Fig. 5.35). As
erythrocytes lack mitochondria, this process also produces excess lactic acid. Regulation of glycolysis in erythrocytes is particularly susceptible to changes in
pH.     Increased     acidity     inhibits     the     glycolysis     initiator     enzyme
(phosphofructokinase-1) as well as certain other allosteric enzymes: pyruvate kinase
and hexokinase. Lower efficiency of pyruvate kinase, as compared to
phosphofructokinase-1, causes more metabolic intermediates to flow through the

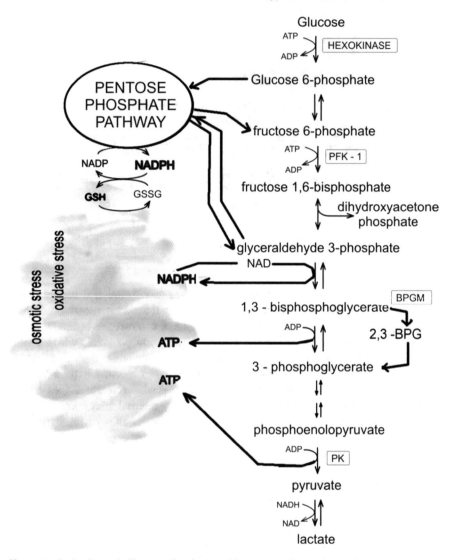

**Fig. 5.35** Red cell metabolism standing in opposition to oxidative and osmotic stresses

oxidative component of the pentose-phosphate cycle. In addition, lower pH also reduces the intensity of the Rapaport-Leubering cycle by inhibiting biphosphoglycerate mutase. Together, these phenomena result in increased production of reducing equivalents, protecting the cell from internal oxidation. This effect is particularly important given the relative abundance of oxygen in erythrocytes traveling from the lungs to other tissues. It ensures that iron is maintained at +2 oxidation, prevents protein clumping, and reduces damage to biological membranes.

It should be noted that old erythrocytes exhibit slightly lower internal pH compared to younger cells.

The erythrocyte fulfills a rigorous biological program which does not provide for substantial freedom of action. It can be compared to a biological "robot" which does not necessarily have to be called alive. Nevertheless, the erythrocyte can also be said to possess some properties of life, most notably a "death trigger" (Fig. 5.35). The abrupt cessation of biological function proves that cell death is not simply a result of stacking internal damage, but instead results from a timer-like mechanism which leads to programmed suicide. The need for such a mechanism is dictated by the necessity of replacing senescent cells and controlling the degradation of cells following their death. In humans, approximately 200 billion erythrocytes die each day, releasing a substantial quantity of hemoglobin which could easily overwhelm the organism's compensatory mechanisms. Controlling the degradation of erythrocytes is therefore very important. Much like apoptosis, the process is biologically programmed. In order to distinguish it from the death of eukaryotic cells, it has been called *eryptosis*. It occurs as a result of osmotic or oxidative stress or when the cell cannot produce sufficient energy to power its membrane pumps. Eryptosis is characterized by an increase in the concentration of calcium ions. Only approximately 0.06%–0.4% of erythrocytes suffer accidental (unprogrammed) death each day, regardless of age.

According to the presented criteria, the erythrocyte—unlike viruses—may be called a living entity because it performs its function automatically and is subject to programmed death.

Scientific consensus holds that in eukaryotic cells senescence and death are controlled by a mechanism which involves progressive shortening of telomeres. However, erythrocytes lack chromosomes and must instead rely on a different process. It is known that erythrocytes (including those stored in donor blood) progressively shed fragments of their membranes. This phenomenon is most probably associated with biologically conditioned instabilities in certain areas of the membrane. It becomes more rapid in an acidic environment and seems to correspond to increases in ambient temperature. Damage sustained by the membrane lowers its active area and causes loss of protein microvesicles. After a certain number of cycles spent carrying oxygen from the lungs to other parts of the organism, the erythrocyte is no longer able to perform the required metabolic actions during its brief stay in each of these tissues. The maximum number of lung tissue cycles is similar in most organisms.

The relation between the erythrocyte's lifespan and the weight of the animal is also fairly well known and enables us to derive a mathematical expression for its longevity. Cyclical reductions in the cell's surface area and volume eventually trigger changes which turn the erythrocyte into a target for phages. As the cell sheds fragments of its membrane, it also loses substances protecting it from being absorbed by the reticuloendothelial system in the liver and spleen. The lower the number of microvesicles in the membrane, the higher the concentration of certain integrated proteins, which eventually begin to clump. Such protein aggregates are recognized by immunoglobulins, resulting in the erythrocyte being removed from the bloodstream. Aggregation of integrated proteins also impacts the activity of enzymes participating in the cell's energy management processes. An example is the AE1 protein, whose aggregates inhibit certain glycolytic enzymes in senescent red blood cells. The same protein also facilitates anion transduction—binding to enzymes renders it unable to maintain a proper ion gradient inside the cell. Decreases in surface area affect the overall shape of the cell and—consequently—the distribution of cytoskeletal proteins. A dying erythrocyte is characterized by increased levels of phosphatidylserine in the outer phospholipid membrane layer. This is a result of lowered availability of ATP, which is required by enzymes responsible for moving this phospholipid into the inner membrane layer (flippases). A sudden increase in the concentration of calcium ions, characteristic of eryptosis, activates an enzymatic protein called scramblase, which also contributes to the exposure of phosphatidylserine. Together both processes trigger mechanisms which degrade the aging cell and clear it from the bloodstream.

Inhibition of any component of interlinked metabolic pathways results in metabolic dysregulation which may lead to cell death. This type of process is observed in diseases which lower the expected lifespan of certain cells—hemolytic anemias caused by deficiencies or inactivation of glycolytic enzymes, defective pentose-phosphate cycles, as well as cytoskeletal abnormalities and dysfunction of integral proteins which constitute biological membranes. Figure 5.31 presents a simplified diagram of erythrocyte metabolic pathways.

The mechanism which triggers erythrocyte death is specific to this group of cells, proving the importance and universality of programmed death. It seems that protein microvesicles perform the function of a biological clock. Once triggered, the death process closely resembles apoptosis. Dead cells are removed from the bloodstream by phages. It seems that while automation imparts biological structures with certain properties of life, the true requirement of inclusion in the animate world is the presence of a programmed death mechanism.

We can therefore conclude that a natural entity is alive if it exhibits autonomy as a result of automatism and follows a programmed mechanism of action which includes a timed death trigger.

It is worth noting that living organisms make frequent use of clock-like mechanisms. Such oscillators are usually based on negative feedback loops, but they also involve positive feedback loops in order to ensure suitable signal properties. They exploit specific biological phenomena as a means of measuring time, often with far-reaching consequences. The ubiquitous nature of oscillators found in all types of cells (including bacteria) seems to indicate the reliance of nature on cyclical processes. The frequency of oscillations varies—from one cycle per second to one per day or even one per year. In the human organism, a particularly important task is performed by the circadian rhythm controlled by a so-called master clock, which coordinates the action of other systemic oscillators. Its development is closely linked to the evolution of life on Earth, governed by the day-night cycle. The circadian rhythm works by anticipating changes in activity associated with various parts of the day, which is why the master clock located in the suprachiasmatic nucleus (NCS) is linked to the retina through a dedicated neural pathway.

A different type of biological oscillator can be found inside cells. Intracellular oscillators measure various frequencies (driven, e.g., by changes in the activity of glycolytic enzymes, calcium ion levels, etc.) and exploit feedback loops which link information stored in DNA to protein products. The master circadian clock seems to be affected by cyclical changes in the activity of transcription factors (CLOCK and BMAL-1) which, in turn, induce the transcription of repressors (Per and Cry proteins); however the role of this mechanism is not precisely known. To date research suggests a connection between the action of oscillators and intracellular metabolic pathways. The universality of this phenomenon may indicate the fundamental importance of actively reinforcing certain biological processes, which— being subject to automatic control—exhibit a natural tendency to slow down having reached a preprogrammed level of activity.

Further research may bring an answer to the question whether oscillators are an indispensable component of living organisms and whether their presence can be included among the criteria of life.

## Suggested Reading

Abdel-Nour M, Carneiro LAM, Downey J, Tsalikis J, Outlioua A, Prescott D, Da Costa LS, Hovingh ES, Farahvash A, Gaudet RG, Molinaro R, van Dalen R, Lau CCY, Azimi FC, Escalante NK, Trotman-Grant A, Lee JE, Gray-Owen SD, Divangahi M, Chen JJ, Philpott DJ, Arnoult D, Girardin SE (2019) The heme-regulated inhibitor is a cytosolic sensor of protein misfolding that controls innate immune signaling. Science 365(6448):eaaw4144. https://doi.org/10.1126/science.aaw4144

Auderset A, Moretti S, Taphorn B, Ebner PR, Kast E, Wang XT, Schiebel R, Sigman DM, Haug GH, Martínez-García A (2022) Enhanced ocean oxygenation during Cenozoic warm periods. Nature 609(7925):77–82. https://doi.org/10.1038/s41586-022-05017-0

Bae T, Fasching L, Wang Y, Shin JH, Suvakov M, Jang Y, Norton S, Dias C, Mariani J, Jourdon A, Wu F, Panda A, Pattni R, Chahine Y, Yeh R, Roberts RC, Huttner A, Kleinman JE, Hyde TM, Straub RE, Walsh CA, Brain Somatic Mosaicism Network§, Urban AE, Leckman JF, Weinberger DR, Vaccarino FM, Abyzov A (2022) Analysis of somatic mutations in 131 human brains reveals aging-associated hypermutability. Science 377(6605):511–517. https://doi.org/10.1126/science.abm6222

Blackburn EH, Epel ES, Lin J (2015) Human telomere biology: a contributory and interactive factor in aging, disease risks, and protection. Science 350(6265):1193–1198. https://doi.org/10.1126/science.aab3389

Boettcher S, Miller PG, Sharma R, McConkey M, Leventhal M, Krivtsov AV, Giacomelli AO, Wong W, Kim J, Chao S, Kurppa KJ, Yang X, Milenkowic K, Piccioni F, Root DE, Rücker FG, Flamand Y, Neuberg D, Lindsley RC, Jänne PA, Hahn WC, Jacks T, Döhner H, Armstrong SA, Ebert BL (2019) A dominant-negative effect drives selection of TP53 missense mutations in myeloid malignancies. Science 365(6453):599–604. https://doi.org/10.1126/science.aax3649

Brown SA, Sato M (2020) Marching to another clock. Science 367(6479):740–741. https://doi.org/10.1126/science.aba5336

Burns KH (2022) Repetitive DNA in disease. Science 376(6591):353–354. https://doi.org/10.1126/science.abl7399

Carroll MP, Spivak JL, McMahon M, Weich N, Rapp UR, May WS (1991) Erythropoietin induces Raf-1 activation and Raf-1 is required for erythropoietin-mediated proliferation. J Biol Chem 266:14964–14969

Cartier A, Hla T (2019) Sphingosine 1-phosphate: lipid signaling in pathology and therapy. Science 366(6463):eaar5551. https://doi.org/10.1126/science.aar5551

Castillo-Davis CI, Mekhedov SL, Hartl DL, Koonin EV (2002) Selection for short introns in highly expressed genes. Nature Genetics. 31:415–418

Chen W, Chen L, Zhang X, Yang N, Guo J, Wang M, Ji S, Zhao X, Yin P, Cai L, Xu J, Zhang L, Han Y, Xiao Y, Xu G, Wang Y, Wang S, Wu S, Yang F, Jackson D, Cheng J, Chen S, Sun C, Qin F, Tian F, Fernie AR, Li J, Yan J, Yang X (2022) Convergent selection of a WD40 protein that enhances grain yield in maize and rice. Science 375(6587):eabg7985. https://doi.org/10.1126/science.abg7985

Cipponi A, Goode DL, Bedo J, McCabe MJ, Marina Pajic M, David R, Croucher DR, Alvaro Gonzalez Rajal AG, Simon R, Junankar SR, Darren N, Saunders DN, Pavel Lobachevsky P, Anthony T, Papenfuss AT, Danielle Nessem D, Max Nobis M, Sean C, Warren SC, Paul Timpson P, Mark Cowley M, Ana C, Vargas AC, Min R, Qiu MR, Daniele G, Generali DG, Shivakumar Keerthikumar S, Uyen Nguyen U, Niall M, Corcoran NM, Georgina V, Long GV, Jean-Yves Blay J-Y, David M, Thomas DM (2020) MTOR signaling orchestrates stress-induced mutagenesis, facilitating adaptive evolution in cancer. Science 368(6495):1127–1131. https://doi.org/10.1126/science.aau8768

Cochran JR, Aivazian D, Cameron TO, Stern LJ (2001) Receptor clustering and transmembrane signalling in T cells. TIBS 26:304–310

Cristiano S, Leal A, Phallen J, Fiksel J, Adleff V, Bruhm DC, Jensen SØ, Medina JE, Hruban C, White JR, Palsgrove DN, Niknafs N, Anagnostou V, Forde P, Naidoo J, Marrone K, Brahmer J, Woodward BD, Husain H, van Rooijen KL, Ørntoft MW, Madsen AH, van de Velde CJH, Verheij M, Cats A, Punt CJA, Vink GR, van Grieken NCT, Koopman M, Fijneman RJA, Johansen JS, Nielsen HJ, Meijer GA, Andersen CL, Scharpf RB, Velculescu VE (2019) Genome-wide cell-free DNA fragmentation in patients with cancer. Nature 570(7761): 385–389. https://doi.org/10.1038/s41586-019-1272-6

Crosby D, Bhatia S, Brindle KM, Coussens LM, Dive C, Emberton M, Esener S, Fitzgerald RC, Gambhir SS, Kuhn P, Emberton M, Rebbeck TR, Balasubramanian S (2022) Early detection of cancer. Nature 379:1244

Desdín-Micó G, Soto-Heredero G, Aranda JF, Oller J, Carrasco E, Gabandé-Rodríguez E, Blanco EM, Alfranca A, Cussó L, Desco M, Ibañez B, Gortazar AR, Fernández-Marcos P, Navarro MN, Hernaez B, Alcamí A, Baixauli F, Mittelbrunn M (2020) T cells with dysfunctional mitochondria induce multimorbidity and premature senescence. Science 368(6497): 1371–1376. https://doi.org/10.1126/science.aax0860

Dudek M, Frendo J, Koj A (1980) Subcellular compartmentation of rhodanese and 3-mercaptopyruvate sulphurtransferase in the liver of some verterbrate species Comp. Bioche Physiol 65B:383–386

Eisenberg E, Levanon EY (2003) Human housekeeping genes are compact. Trends in Genetics. 19: 362–365

Eliasmith C, Stewart TC, Choo X, Bekolay T, DeWolf T, Tang C, Rasmussen D (2012) A large-scale model of the functioning brain. Science 338:1202–1205

Esposito E, Li W, Mandeville T, Park JH, Şencan I, Guo S, Shi J, Lan J, Lee J, Hayakawa K, Sakadžić S, Ji X, Lo EH (2020) Potential circadian effects on translational failure for neuroprotection. Nature 582(7812):395–398. https://doi.org/10.1038/s41586-020-2348-z

Esterházy D, Canesso MCC, Mesin L, Muller PA, de Castro TBR, Lockhart A, ElJalby M, Faria AMC, Mucida D (2019) Compartmentalized gut lymph node drainage dictates adaptive immune responses. Nature 569(7754):126–130. https://doi.org/10.1038/s41586-019-1125-3

Fields S, Johnston M (2005) Whither model organism research? Science 307:1885–1888

Flavahan WA, Drier Y, Johnstone SE, Hemming ML, Tarjan DR, Hegazi E, Shareef SJ, Javed NM, Raut CP, Eschle BK, Gokhale PC, Hornick JL, Sicinska ET, Demetri GD, Bernstein BE (2019) Altered chromosomal topology drives oncogenic programs in SDH-deficient GISTs. Nature 575(7781):229–233. https://doi.org/10.1038/s41586-019-1668-3

Föller M, Huber SM, Lang F (2008) Erythrocyte programmed cell death. IUBMB Life 60:661–668

Frottin F, Schueder F, Tiwary S, Gupta R, Körner R, Schlichthaerle T, Cox J, Jungmann R, Hartl FU, Hipp MS (2019) The nucleolus functions as a phase-separated protein quality control compartment. Science 365(6451):342–347. https://doi.org/10.1126/science.aaw9157

García-Bayona L, Comstock LE (2018) Bacterial antagonism in host-associated microbial communities. Science 361(6408):eaat2456. https://doi.org/10.1126/science.aat2456

Gilette MU, Sejnowski TJ (2005) Biological clocks coordinately keep life on time. Science 309: 1196–1198

Ginsberg HN, Mani A (2022) Complex regulation of fatty liver disease. Science 376(6590): 247–248. https://doi.org/10.1126/science.abp8276

Giumerá R, Amaral LAN (2005) Functional cartography of complex metabolic networks. Nature 433:895–900

Glass DS, Alon U (2018) Programming cells and tissues. Science 361(6408):1199–1200. https://doi.org/10.1126/science.aav2497

Goldbeter A (2002) Computational approaches to cellular rhythms. Nature 420:238–245

Goodell MA, Rando TA (2015) Stem cells and healthy aging. Science 350(6265):1199–1204. https://doi.org/10.1126/science.aab3388

Han X, Zhou Z, Fei L, Sun H, Wang R, Chen Y, Chen H, Wang J, Tang H, Ge W, Zhou Y, Ye F, Jiang M, Wu J, Xiao Y, Jia X, Zhang T, Ma X, Zhang Q, Bai X, Lai S, Yu C, Zhu L, Lin R, Gao Y, Wang M, Wu Y, Zhang J, Zhan R, Zhu S, Hu H, Wang C, Chen M, Huang H, Liang T, Chen J, Wang W, Zhang D, Guo G (2020) Construction of a human cell landscape at single-cell level. Nature 581(7808):303–309. https://doi.org/10.1038/s41586-020-2157-4

Haynes CM (2015) Cell biology: surviving import failure. Nature 524(7566):419–420. https://doi.org/10.1038/nature14644

Helmink BA, Reddy SM, Gao J, Zhang S, Basar R, Thakur R, Yizhak K, Sade-Feldman M, Blando J, Han G, Gopalakrishnan V, Xi Y, Zhao H, Amaria RN, Tawbi HA, Cogdill AP, Liu W, LeBleu VS, Kugeratski FG, Patel S, Davies MA, Hwu P, Lee JE, Gershenwald JE, Lucci A, Arora R, Woodman S, Keung EZ, Gaudreau PO, Reuben A, Spencer CN, Burton EM, Haydu LE, Lazar AJ, Zapassodi R, Hudgens CW, Ledesma DA, Ong S, Bailey M, Warren S, Rao D, Krijgsman O, Rozeman EA, Peeper D, Blank CU, Schumacher TN, Butterfield LH, Zelazowska MA, McBride KM, Kalluri R, Allison J, Petitprez F, Fridman WH, Sautès-Fridman C, Hacohen N, Rezvani K, Sharma P, Tetzlaff MT, Wang L, Wargo JA (2020) B cells and tertiary lymphoid structures promote immunotherapy response. Nature 577(7791): 549–555. https://doi.org/10.1038/s41586-019-1922-8

Joh HK, Lee DH, Hur J, Nimptsch K, Chang Y, Joung H, Zhang X, Rezende LFM, Lee JE, Ng K, Yuan C, Tabung FK, Meyerhardt JA, Chan AT, Pischon T, Song M, Fuchs CS, Willett WC, Cao Y, Ogino S, Giovannucci E, Wu K (2021) simple sugar and sugar-sweetened beverage intake during adolescence and risk of colorectal cancer precursors. Gastroenterology 161(1): 128–142.e20. https://doi.org/10.1053/j.gastro.2021.03.028

Kim B, Kanai MI, Oh Y, Kyung M, Kim EK, Jang IH, Lee JH, Kim SG, Suh GSB, Lee WJ (2021) Response of the microbiome-gut-brain axis in Drosophila to amino acid deficit. Nature 593(7860):570–574. https://doi.org/10.1038/s41586-021-03522-2

Kiraly DD (2019) Gut microbes regulate neurons to help mice forget their fear. Nature 574(7779): 488–489. https://doi.org/10.1038/d41586-019-03114-1

Klein IA, Boija A, Afeyan LK, Hawken SW, Fan M, Dall'Agnese A, Oksuz O, Henninger JE, Shrinivas K, Sabari BR, Sagi I, Clark VE, Platt JM, Kar M, McCall PM, Zamudio AV, Manteiga JC, Coffey EL, Li CH, Hannett NM, Guo YE, Decker TM, Lee TI, Zhang T, Weng JK, Taatjes DJ, Chakraborty A, Sharp PA, Chang YT, Hyman AA, Gray NS, Young RA (2020) Partitioning of cancer therapeutics in nuclear condensates. Science 368(6497):1386–1392. https://doi.org/10.1126/science.aaz4427

Krndija D, El Marjou F, Guirao B, Richon S, Leroy O, Bellaiche Y, Hannezo E, Matic Vignjevic D (2019) Active cell migration is critical for steady-state epithelial turnover in the gut. Science 365(6454):705–710. https://doi.org/10.1126/science.aau3429

Lang KS, Lang PA, Bauer C, Duranton C, Wieder T, Huber SM, Lang F (2005) Mechanisms of suicidal erythrocyte death. Cell Physiol Biochem 15:195–202

Lee JH, Liu R, Li J, Zhang C, Wang Y, Cai Q, Qian X, Xia Y, Zheng Y, Piao Y, Chen Q, de Groot JF, Jiang T, Lu Z (2017) Stabilization of phosphofructokinase 1 platelet isoform by AKT promotes tumorigenesis. Nat Commun 8(1):949. https://doi.org/10.1038/s41467-017-00906-9

Lee J-Y, Tsolis RM, Bäumler AJ (2022) The microbiome and gut homeostasis. Science 377(6601): eabp9960. https://doi.org/10.1126/science.abp9960

Lee SB, Frattini V, Bansal M, Castano AM, Sherman D, Hutchinson K, Bruce JN, Califano A, Liu G, Cardozo T, Iavarone A, Lasorella A (2016) An ID2-dependent mechanism for VHL inactivation in cancer. Nature 529(7585):172–177. https://doi.org/10.1038/nature16475

Lee-Six H, Olafsson S, Ellis P, Osborne RJ, Sanders MA, Moore L, Georgakopoulos N, Torrente F, Noorani A, Goddard M, Robinson P, Coorens THH, O'Neill L, Alder C, Wang J, Fitzgerald RC, Zilbauer M, Coleman N, Saeb-Parsy K, Martincorena I, Campbell PJ, Stratton MR (2019) The landscape of somatic mutation in normal colorectal epithelial cells. Nature 574(7779):532–537. https://doi.org/10.1038/s41586-019-1672-7

Levescot A, Cerf-Bensussan N (2022) Regulatory CD8 $^+$ T cells suppress disease. Science 376(6590):243–244. https://doi.org/10.1126/science.abp8243

Li XV, Leonardi I, Putzel GG, Semon A, Fiers WD, Kusakabe T, Lin W-Y, Gao IH, Doron I, Gutierrez-Guerrero A, DeCelie MB, Carriche GM, Mesko M, Yang C, Naglik JR, Hube B, Scherl EJ, Iliev ID (2022) Immune regulation by fungal strain diversity in inflammatory bowel disease. Nature 603(7902):672–678. https://doi.org/10.1038/s41586-022-04502-w

Lu Q, Stappenbeck TS (2022) Local barriers configure systemic communications between the host and microbiota. Science 376(6596):950–955. https://doi.org/10.1126/science.abo2366

Maley CC, Shibata D (2019) Cancer cell evolution through the ages. Science 365(6452):440–441. https://doi.org/10.1126/science.aay2859

Mao C, Liu X, Zhang Y, Lei G, Yan Y, Lee H, Koppula P, Wu S, Zhuang L, Fang B, Poyurovsky MV, Olszewski K, Gan B (2021) DHODH-mediated ferroptosis defence is a targetable vulnerability in cancer. Nature 593(7860):586–590. https://doi.org/10.1038/s41586-021-03539-7

Margueron R, Reinberg D (2011) The polycomb complex PRC2 and its mark in life. Nature 467: 343–349

McDonald B, McCoy KD (2019) Maternal microbiota in pregnancy and early life. Science 365(6457):984–985. https://doi.org/10.1126/science.aay0618

Mihalcescu I, Hsing W, Leibler S (2004) Resilient circadian oscillator revealed in individual cyanobacteria. Nature 430:81–85

Nowosad CR, Mesin L, Castro TBR, Wichmann C, Donaldson GP, Araki T, Schiepers A, Lockhart AAK, Bilate AM, Mucida D, Victora GD (2020) Tunable dynamics of B cell selection in gut germinal centres. Nature 588(7837):321–326. https://doi.org/10.1038/s41586-020-2865-9

Ochman H, Raghavan R (2009) Excavating the functional landscape of bacterial cells. Science 326: 1200–1201

Olin-Sandoval V, Yu JSL, Miller-Fleming L, Alam MT, Kamrad S, Correia-Melo C, Haas R, Segal J, Peña Navarro DA, Herrera-Dominguez L, Méndez-Lucio O, Vowinckel J, Mülleder M, Ralser M (2019) Lysine harvesting is an antioxidant strategy and triggers underground polyamine metabolism. Nature 572(7768):249–253. https://doi.org/10.1038/s41586-019-1442-6

O'Toole PW, Jeffery IB (2015) Gut microbiota and aging. Science 350(6265):1214–1215. https://doi.org/10.1126/science.aac8469

Ozbudak EM, Thattai M, Lim HN, Shraiman BI, van Oudenaarden A (2004) Multistability in the lactose utilization network of Escherichia coli. Nature 427:737–740

Palmer JD, Foster KR (2022) Bacterial species rarely work together. Science 376(6593):581–582. https://doi.org/10.1126/science.abn5093

Papin JA, Reed JL, Palsson BO (2004) Hierarchical thinking in network biology: the unbiased modularization of biochemical networks. TIBS 29:641–647

Patel S, Churchill GC, Galione A (2001) Coordination of $Ca^{2+}$ signalling by NAADP. TIBS 26: 482–489

Petljak M, Dananberg A, Chu K, Bergstrom EN, Striepen J, von Morgen P, Chen Y, Shah H, Sale JE, Alexandrov LB, Stratton MR, Maciejowski J (2022) Mechanisms of APOBEC3 mutagenesis in human cancer cells. Nature 607(7920):799–807. https://doi.org/10.1038/s41586-022-04972-y

Petsko GA, Small SA (2022) Elucidating the causes of neurodegeneration. Science 377(6601): 31–32. https://doi.org/10.1126/science.adc9969

Pierre P (2019) Integrating stress responses and immunity. Science 365(6448):28–29. https://doi.org/10.1126/science.aay0987

Priestley P, Baber J, Lolkema MP, Steeghs N, Bruijn E, Shale C, Duyvesteyn K, Haidari S, van Hoeck A, Onstenk W, Roepman P, Voda M, Bloemendal HJ, Tjan-Heijnen VCG, van Herpen CML, Labots M, Witteveen PO, Smit EF, Sleijfer S, Voest EE, Cuppen E (2019) Pan-cancer whole-genome analyses of metastatic solid tumours. Nature 575(7781):210–216. https://doi.org/10.1038/s41586-019-1689-y

Raab M, Gentili M, de Belly H, Thiam HR, Vargas P, Jimenez AJ, Lautenschlaeger F, Voituriez R, Lennon-Duménil AM, Manel N, Piel M (2016) ESCRT III repairs nuclear envelope ruptures during cell migration to limit DNA damage and cell death. Science 352(6283):359–362. https://doi.org/10.1126/science.aad7611

Ray S, Valekunja UK, Stangherlin A, Howell SA, Snijders AP, Damodaran G, Reddy AB (2020) Circadian rhythms in the absence of the clock gene Bmal1. Science 367(6479):800–806. https://doi.org/10.1126/science.aaw7365

Reddien PW (2019) The cells of regeneration. Science 365(6451):314–316. https://doi.org/10.1126/science.aay3660

Rivenbark AG, Strahl BD (2007) Unlocking cell fate. Science 318:403–404

Rosshart SP, Herz J, Vassallo BG, Hunter A, Wall MK, Badger JH, McCulloch JA, Anastasakis DG, Sarshad AA, Leonardi I, Collins N, Blatter JA, Han S-J, Tamoutounour S, Potapova S, St Claire MBF, Yuan W, Sen SK, Dreier MS, Hild B, Hafner M, Wang D, Iliev ID, Belkaid Y, Trinchieri G, Rehermann B (2019) Laboratory mice born to wild mice have natural microbiota and model human immune responses. Science 365(6452):eaaw4361. https://doi.org/10.1126/science.aaw4361

Sandstrom A, Mitchell PS, Goers L, Mu EW, Lesser CF, Vance RE (2019) Functional degradation: a mechanism of NLRP1 inflammasome activation by diverse pathogen enzymes. Science 364(6435):eaau1330. https://doi.org/10.1126/science.aau1330

Schaum N, Lehallier B, Hahn O, Pálovics R, Hosseinzadeh S, Lee SE, Sit R, Lee DP, Losada PM, Zardeneta ME, Fehlmann T, Webber JT, McGeever A, Calcuttawala K, Zhang H, Berdnik D, Mathur V, Tan W, Zee A, Tan M, Tabula Muris Consortium, Pisco AO, Karkanias J, Neff NF, Keller A, Darmanis S, Quake SR, Wyss-Coray T (2020) Ageing hallmarks exhibit organ-specific temporal signatures. Nature 583(7817):596–602. https://doi.org/10.1038/s41586-020-2499-y

Schlieben LD, Prokisch H (2020) The dimensions of primary mitochondrial disorders. Front Cell Dev Biol 8:600079. https://doi.org/10.3389/fcell.2020.600079

Schluter J, Peled JU, Taylor BP, Markey KA, Smith M, Taur Y, Niehus R, Staffas A, Dai A, Fontana E, Amoretti LA, Wright RJ, Morjaria S, Fenelus M, Pessin MS, Chao NJ, Lew M, Bohannon L, Bush A, Sung AD, Hohl TM, Perales M-A, van den Brink MRM, Xavier JB (2020) The gut microbiota is associated with immune cell dynamics in humans. Nature 588(7837):303–307. https://doi.org/10.1038/s41586-020-2971-8

Schwartz MA, Vestweber D, Simons M (2018) A unifying concept in vascular health and disease. Science 360(6386):270–271. https://doi.org/10.1126/science.aat3470

Scott JD, Pawson T (2000) Cell communication: the inside story. Sci Am 282(6):72–79

Sonnenburg ED, Smits SA, Tikhonov M, Higginbottom SK, Wingreen NS, Sonnenburg JL (2016) Diet-induced extinctions in the gut microbiota compound over generations. Nature 529(7585): 212–215. https://doi.org/10.1038/nature16504

Strecker J, Ladha A, Gardner Z, Schmid-Burgk JL, Makarova KS, Koonin EV, Zhang F (2019) RNA-guided DNA insertion with CRISPR-associated transposases. Science 365(6448):48–53. https://doi.org/10.1126/science.aax9181

Sun H, Tonks NK (1994) The coordinated action of protein tyrosine phosphatases and kinases in cell signaling. TIBS 19:480–485

Szüts D (2022) A fresh look at somatic mutations in cancer. Science 376(6591):351–352. https://doi.org/10.1126/science.abo742

Tabula Muris Consortium (2020) A single-cell transcriptomic atlas characterizes ageing tissues in the mouse. Nature 583(7817):590–595. https://doi.org/10.1038/s41586-020-2496-1

Tokheim CJ, Papadopoulos N, Kinzler KW, Vogelstein B, Karchin R (2016) Evaluating the evaluation of cancer driver genes. Proc Natl Acad Sci U S A 113(50):14330–14335. https://doi.org/10.1073/pnas.1616440113

Tuveson D, Clevers H (2019) Cancer modeling meets human organoid technology. Science 364(6444):952–955. https://doi.org/10.1126/science.aaw6985

Wang H, Nakamura M, Abbott TR, Zhao D, Luo K, Yu C, Nguyen CM, Lo A, Daley TP, La Russa M, Liu Y, Qi LS (2019) CRISPR-mediated live imaging of genome editing and transcription. Science 365(6459):1301–1305. https://doi.org/10.1126/science.aax7852

Watkins TBK, Lim EL, Petkovic M, Elizalde S, Birkbak NJ, Wilson GA, Moore DA, Grönroos E, Rowan A, Dewhurst SM, Demeulemeester J, Dentro SC, Horswell S, Au L, Haase K, Escudero M, Rosenthal R, Bakir MA, Xu H, Litchfield K, Lu WT, Mourikis TP, Dietzen M, Spain L, Cresswell GD, Biswas D, Lamy P, Nordentoft I, Harbst K, Castro-Giner F, Yates LR, Caramia F, Jaulin F, Vicier C, Tomlinson IPM, Brastianos PK, Cho RJ, Bastian BC, Dyrskjøt L, Jönsson GB, Savas P, Loi S, Campbell PJ, Andre F, Luscombe NM, Steeghs N, Tjan-Heijnen VCG, Szallasi Z, Turajlic S, Jamal-Hanjani M, Van Loo P, Bakhoum SF, Schwarz RF, McGranahan N, Swanton C (2020) Pervasive chromosomal instability and karyotype order in tumour evolution. Nature 587(7832):126–132. https://doi.org/10.1038/s41586-020-2698-6

Wolin SL, Maquat LE (2019) Cellular RNA surveillance in health and disease. Science 366(6467):822–827. https://doi.org/10.1126/science.aax2957

Yalcin A, Telang S, Clem B, Chesney J (2009) Regulation of glucose metabolism by 6-phosphofructo-2-kinase/fructose-2,6-bisphosphatases in cancer. Exp Mol Pathol 86(3):174–179. https://doi.org/10.1016/j.yexmp.2009.01.003

Yixian Cui Y, Smriti Parashar S, Muhammad Zahoor M, Patrick G, Needham PG, Muriel Mari M, Ming Zhu M, Shuliang Chen S, Hsuan-Chung Ho H-C, Fulvio Reggiori F, Hesso Farhan H, Jeffrey L, Brodsky JL, Susan Ferro-Novick S (2019) A COPII subunit acts with an autophagy receptor to target endoplasmic reticulum for degradation. Science 365(6448):53–60. https://doi.org/10.1126/science.aau9263

Yuan J, Chang SY, Yin SG, Liu ZY, Cheng X, Liu XJ, Jiang Q, Gao G, Lin DY, Kang XL, Ye SW, Chen Z, Yin JA, Hao P, Jiang L, Cai SQ (2020) Two conserved epigenetic regulators prevent healthy ageing. Nature 579(7797):118–122. https://doi.org/10.1038/s41586-020-2037-y

Yu J, Vodyanik MA, Smuga-Otto K, Antosiewicz-Bourget J, Frane JL, Tian S, Nie J, Jonsdottir GA, Ruotti V, Stewart R, Slukvin II, Thomson JA (2007) Induced pluripotent stem cell lines derived from human somatic cells. Science 318:1917–1922

Zheng W, Zhao W, Wu M, Song X, Caro F, Sun X, Gazzaniga F, Stefanetti G, Oh S, Mekalanos JJ, Kasper DL (2020) Microbiota-targeted maternal antibodies protect neonates from enteric infection. Nature 577(7791):543–548. https://doi.org/10.1038/s41586-019-1898-4

Printed in the United States
by Baker & Taylor Publisher Services